A-LEVEL GEOGRAPHY TOPIC MASTER

COASTAL LANDSCAPES

Series editor
Simon Oakes

Peter Stiff

HODDER
EDUCATION
AN HACHETTE UK COMPANY

For Dominic and Nathaniel – may they come to enjoy, respect and care for coastal landscapes and theirinhabitants.

Acknowledgements including photo credits can be found on page 228.

Orders: please contact Hachette UK Distribution, Hely Hutchinson Centre, Milton Road, Didcot, Oxfordshire, OX11 7HH. Telephone: +44 (0)1235 827827. Email education@hachette.co.uk. Lines are open from 9 a.m. to 5 p.m., Monday to Friday. You can also order through our website: www.hoddereducation.co.uk

ISBN: 9781510434622

© Peter Stiff 2018
First published in 2018 by
Hodder Education,
An Hachette UK Company
Carmelite House
50 Victoria Embankment
London EC4Y 0DZ
www.hoddereducation.co.uk

Impression number   10 9 8 7 6 5 4 3

Year   2024

Cover photo © Larry Geddis/Alamy Stock Photo

Illustrations by Barking Dog and Aptara Inc.

Typeset in India by Aptara Inc.

Printed by CPI Group (UK) Ltd, Croydon CR0 4YY

A catalogue record for this title is available from the British Library.

Contents

Introduction

The coast is a dynamic environment, its unique combination of sea, atmosphere and land creating both dramatic and subtle landforms and landscapes. Reforms of the A-level curriculum reinforced the study of physical geography, something this book aims to support and encourage, using examples and case studies drawn from around the world. The need to develop detailed knowledge and understanding of coastal systems and processes is made urgent by the challenges posed as sea level rises in response to global warming. The increasing threat of erosion and flooding means that sustainable management is required across the development continuum. This book also covers the value coastal landscapes have for settlement, transport, trade, tourism and mineral extraction.

The A-level Geography Topic Master series

The books in this series are designed to support learners who aspire to reach the highest grades. To do so requires more than rote learning. Only around one-third of available marks in an A-level Geography examination are allocated to the recall of knowledge (*assessment objective 1*, or *AO1*). A greater proportion is reserved for higher-order cognitive tasks, including the **analysis**, **interpretation** and **evaluation** of geographic ideas and information (*assessment objective 2*, or *AO2*). Therefore, the material in this book has been purposely written and presented in ways which encourage active reading, reflection and critical thinking. The overarching aim is to help you develop the analytical and evaluative 'geo-capabilities' needed for examination success. Opportunities to practise and develop **data manipulation skills** are also embedded throughout the text (supporting *assessment objective 3*, or *AO3*).

All *Geography Topic Master* books prompt students constantly to 'think geographically'. In practice this can mean learning how to seamlessly integrate **geographic concepts** – including place, scale, interdependency, causality and inequality – into the way we think, argue and write. The books also take every opportunity to establish **synoptic links** (this means making 'bridging' connections between themes and topics). Frequent page-referencing is used to create links between different chapters and sub-topics. Additionally, numerous connections have been highlighted between *Coastal landscapes* and other Geography topics, such as *Changing places*, *Global systems* or *Water and carbon cycles*.

Using this book

The book may be read from cover to cover since there is a logical progression between chapters (each of which is divided into four sections). On the other hand, a chapter may be read independently whenever required as part of your school's scheme of work for this topic. A common set of features are used in each chapter:

- *Aims* establish the four main points (and sections) of each chapter.
- *Key concepts* are important ideas relating either to the discipline of Geography as a whole or more specifically to the study of coastal landscapes.
- *Contemporary case studies* apply geographical ideas, theories and concepts to real-world local contexts such as risks from coastal flooding to London and The Maldives and issues involved in managing the coast at a time of rising sea level (the Exe estuary). *Located examples* offer contextual detail but in less depth.
- *Analysis and interpretation* features help you develop the geographic skills and capabilities needed for the application of knowledge and understanding (AO2), data manipulation (AO3) and, ultimately, exam success.
- *Evaluating the issue* brings each chapter to a close by discussing a key coastal landscapes issue (typically involving competing perspectives and views).
- Also included at the end of each chapter are the *Chapter summary*, *Refresher questions*, *Discussion activities*, *Fieldwork focus* (supporting the independent investigation) and selected *Further reading*.

Coastal system dynamics – flows of energy

There is a great variety of landforms and landscapes located along the world's million or so kilometres of coastline. How these features have developed and continue to evolve is the result of the interaction of processes powered by various sources of energy. This chapter:

- analyses coasts as open energy systems
- investigates energy inputs: waves, tides and currents
- explores feedback and equilibrium in coastal systems
- assesses the relative importance of energy sources in the coastal system.

KEY CONCEPTS

Systems – groups of related components. In physical geography, they tend to be 'open', that is having both inputs and outputs. The coast is such a system with both energy and materials, such as sediment, flowing through it. Systems operate across all scales, from a small cove right the way up to the global oceanic system.

Equilibrium – the state of balance within a system. Any stretch of coastline can be investigated in terms of its equilibrium. Some locations might receive an increased level of wave energy due to a fierce storm and this could affect the intensity of wave erosion. In turn an increase in erosion could lead to a dramatic change in a cliff, leading to its collapse.

Feedback – an automatic response to change within a system. Positive feedback leads to further change, for example an increase in wave energy could scour the seabed, deepening the water and allowing yet more wave energy to enter the location, further increasing wave energy. Negative feedback reduces the effect of change, for example a decrease in wave energy entering the coastal zone could lead to deposition of sediment offshore, further reducing wave energy as more wave energy is lost to friction.

Threshold – critical 'tipping points' in a system. For example, a single wave can be considered as a system. As it approaches the shore, wave height increases until the threshold is reached at which the water cannot be supported and the wave breaks, resulting in a forward rush of water.

① Coasts as open systems

▶ *How do inputs, stores and processes and outputs operate together to form the coastal system?*

An open system is one in which energy and materials move across its boundaries. Closed systems are those where only flows of energy occur across their boundaries.

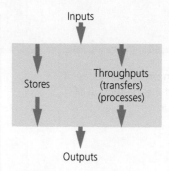

Inputs

Stores

Throughputs
(transfers)
(processes)

Outputs

▲ **Figure 1.1** An open system

All such open systems include:

- inputs, e.g. energy, sediment
- stores or components, e.g. sediment, water
- processes, e.g. weathering, erosion, mass movement
- throughput, e.g. energy, sediment, water
- outputs, e.g. energy, sediment.

The application of systems when studying coasts has much potential as it focuses on knowledge and understanding of flows of energy and materials. These combine to create specific landforms such as cliffs and beaches. However, defining system boundaries is problematic. Can sections of a coast be sensibly divided up, and where does one section finish and another start? Where reasonable boundaries are established, it soon becomes apparent that a hierarchy of systems exists. For example, it is possible to sub-divide a sand dune system into different types of dunes – embryo, fore, yellow, grey. In turn, each individual dune ridge consists of two slopes or faces and a crest.

Using systems is not without its difficulties but it helps identify the various flows and components and understand how landforms and landscapes are created and interact.

Defining the coast

Coasts are where sea, land and atmosphere meet. This unique combination may seem to offer a simple location to study but the reality is complex. Where tall cliffs plunge directly into the sea the boundary between land and sea appears clear. Other locations, such as extensive, low-lying estuaries, seem at times to be **terrestrial** but can change over the course of a few hours into a marine environment as the tide sweeps in.

The most useful context for a study of coasts is to think in terms of a coastal zone. This zone extends between the inland limit of marine influence – for example where sea spray can be blown – and the seaward limit of the land's influence, such as how far river water extends out to sea as an identifiable flow. However, not to be excluded are the marine processes that extend beyond such a zone, such as the circulation of sediment in an offshore cell.

The term **littoral** can also be applied to coasts. However, there is no one single definition of coasts and the use of definitions varies. It can, however, be helpful to describe some spatial boundaries of the coastal zone and to appreciate that sub-divisions can exist (Figure 1.2).

> 🔑 **KEY TERMS**
>
> **Littoral** The environment between the highest and lowest levels that tides reach.
>
> **Quaternary** The most recent geological period, existing for the past approximately 2.5 million years.

(a) Low energy coast

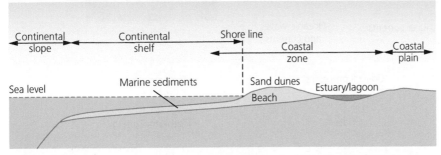

(b) High energy coast

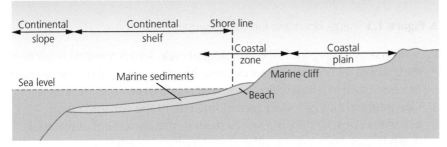

▲ **Figure 1.2** Spatial sub-divisions at the coast

Not all the elements in each type of coast will necessarily be present in all locations. A lagoon does not exist along every low-lying coast as it does, for example, at Slapton, Devon. The impact of the advance and retreat of glaciers and ice sheets over the **Quaternary** has had a very significant role in defining the coastal zone. During this period, sea level has risen and fallen over 100 metres vertically many times. This has left a legacy of coastal processes and landforms both further inland and out to sea than the present-day coastline, something discussed in greater detail in Chapter 5.

The input of energy into the coastal system

Three sources of energy drive the coastal system:

1 Solar energy
2 Gravitational energy
3 Geothermal energy.

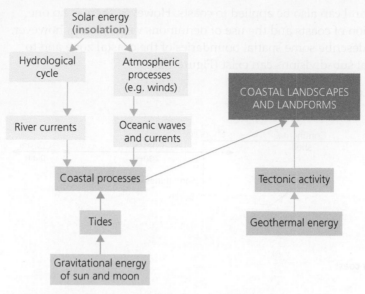

▲ **Figure 1.3** Energy inputs into the coastal system

1 Solar energy – this powers the hydrological cycle, which transfers water from the land into the coastal zone, for example. It is also responsible for atmospheric processes such as winds which, in turn, generate waves and currents.

2 Gravitational energy – the gravitational pull of the sun and moon generates tides. Gravity is a key element in the movement of material down slopes, such as rock fall and underwater landslides.

3 Geothermal energy – responsible for tectonic activity that can cause uplift or submergence along a coast.

KEY TERMS

Insolation Incoming solar radiation. Its intensity per unit area varies with latitude being highest around the Equator and lowest at the Poles.

Archipelago A closely grouped cluster of islands.

It is important to recognise that energy inputs operate over very different time scales. At the geological scale (millions of years), geothermal energy can cause the convergence and subduction of plates along coasts. The southern coastline of the Indonesian archipelago and much of the west coast of South America receive such inputs of energy. Still over long time periods, but measured in hundreds of thousands of years, is the drowning of river valleys to create estuaries following a rise in sea level. At the other end of the time scale, winds and waves can be generated and die away in the matter of a few hours. An individual wave may take barely a minute to break on a beach.

Energy allows 'work' to be undertaken. As it flows through the coastal zone, ecosystems function, sediments can be transported by water and wind, and waves break against a cliff face causing particles to be broken off. Changes in the energy available in the coastal zones changes the 'work' that can be achieved and in turn this affects the landforms and landscapes that develop.

▲ **Figure 1.4** Volcanoes along the Indonesian coastline

Energy inputs – waves, tides and currents

▶ *In what ways is energy brought into the coastal system by waves, tides and currents?*

Waves

All coasts are affected by waves to some extent but there are considerable contrasts from one location to another in the nature of wave activity. This is the case at all scales from the macro (global) to the micro (local). Waves represent a transfer of energy and are capable of carrying out much work in the coastal zone. This work includes erosion, transport and deposition of materials. Waves are described by their:

- height – the vertical distance between wave crest and wave trough
- length – the horizontal distance between consecutive crests
- wave period – the time taken for consecutive crests to pass a fixed point
- wave steepness – the ratio of wave height to wave length
- wave velocity – the ratio of wave length to wave period.

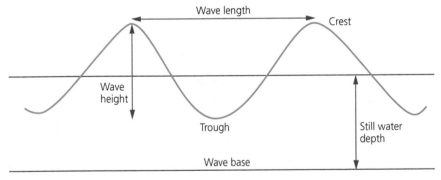

▲ **Figure 1.5** Wave characteristics

Wave height is a useful descriptor as it is commonly taken as an indication of wave energy. Two frequently used formulae for a measure of wave height are:

$$H = 0.031U^2$$

$$H = 0.36 \sqrt{F}$$

where H = wave height, U = wind velocity (ms⁻¹) and F is the fetch (km). 0.031 and 0.36 are empirically derived constants.

The **fetch** is an important factor in the energy transfer from atmosphere to water. The greatest energy transfer occurs when strong winds blow in the same direction over a long distance for a long period of time; this generates the highest waves and so brings significant wave energy into the coastal zone.

 KEY TERM

Fetch The distance of open water in one direction from a coastline, over which the wind can blow.

The energy (E) of a wave in deep water is proportional to the product of the wavelength (L) and the square of the height (H):

$$E \propto LH^2$$

This formula indicates that a small increase in wave height will result in a proportionately much greater increase in energy. Because wave height varies greatly across the course of even a few minutes, a widely used measure of wave height is 'significant wave height'. This is the mean wave height of the highest third of the waves, averaging an instrument record taken over relatively short time periods (a few minutes).

Wave energy at the global scale

On the global scale, it is possible to identify coastlines exposed to different levels of wave energy on the basis of prevailing winds, fetch and aspect. The global distribution of high energy waves reflects the locations of substantial areas of open sea and the pattern of global winds.

Coastlines regularly receiving high wave energy are those of mid-latitude regions, north and south of the equator (Figure 1.6). Here atmospheric storms, circulating round the globe in westerly directions, occur regularly and frequently, generating large waves, that is, greater than five metres.

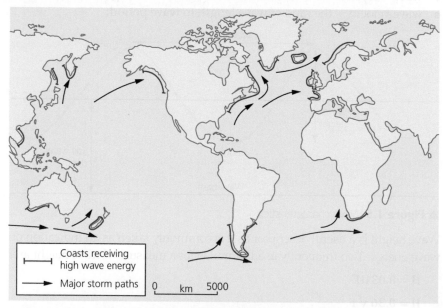

▲ **Figure 1.6** Location of coastlines receiving high wave energy

In the southern hemisphere, the area between 40 and 50 degrees south acquired the name the 'Roaring Forties'. Those on board sailing ships crossing these regions had to contend with severe storms and their accompanying waves. Such hostile conditions had considerable influences on flows of people and goods and the early developments of regional and global economic, social and political systems. Significant wave height around Cape Horn, at the

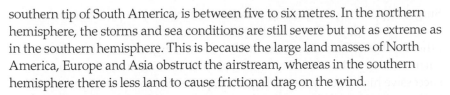
southern tip of South America, is between five to six metres. In the northern hemisphere, the storms and sea conditions are still severe but not as extreme as in the southern hemisphere. This is because the large land masses of North America, Europe and Asia obstruct the airstream, whereas in the southern hemisphere there is less land to cause frictional drag on the wind.

Other coastlines experiencing storm waves include some in the lower, tropical and sub-tropical latitudes such as South-east Asia and areas around the Indian Ocean and Arabian Sea. Here, seasonal storms associated with the monsoons bring increased amounts of wave energy into the coastal zone.

Additionally, coastlines in areas where tropical storms and hurricanes (cyclones/typhoons) are generated receive very high levels of wave energy when these storms make landfall.

Beyond these areas, coastlines exposed to open stretches of water still receive significant inputs of wave energy. Significant wave height for the western British Isles is between four and five metres. Where a sea is essentially enclosed, such as the Mediterranean, relatively low wave energy is experienced. Significant wave height here is between one and two metres. However, the lowest wave energy exists in polar regions where coastal sea ice is an extensive feature, such as Arctic coastlines.

Waves generated well away from land, for example out in the Pacific or Southern Ocean, become swell waves. These travel long distances in relatively regular progression. They are unaffected by local winds and can be seen moving against a local wind flow.

Dominant waves are key in understanding energy inputs from the ocean-atmosphere systems. Often these are the result of prevailing winds, but in some locations locally 'dominant winds' generate locally dominant waves. It is important, therefore, to consider influences at contrasting scales when investigating coastal landforms and landscapes.

Storm surges

Three factors combine to create a storm surge:

- extremely high onshore wind speed
- very low atmospheric pressure
- the shape of the coastline.

The high wind speeds are usually associated with the passage of a tropical storm. If the storm reaches a coastal area at the same time as high tide occurs, sea level rises even higher.

When air in the lower atmosphere is rising, less mass of air presses down on the sea surface, thereby allowing the water to rise well above normal levels. Sea level rises approximately one centimetre for every one millibar fall in air pressure.

Along funnel-shaped coasts where there is a relatively shallow off-shore gradient, surges are more pronounced. Relatively straight coastlines are not as affected.

 KEY TERMS

Swell waves These are generated out at sea and have a long wave period.

Dominant wave The wave which has the greatest influence on a particular stretch of coast.

Storm surge An elevated sea surface near the coast, most often caused by extremely energetic storms such as hurricanes/typhoons/cyclones.

Surges happen regularly along coastlines affected by tropical storms, such as around the Bay of Bengal or the Caribbean. However, surges can occur in other locations. On 5 December 2013, a storm surge affected coastlines surrounding the North Sea and through into the Channel. The two successive high tides were the highest experienced in some 60 years and led to the evacuation of thousands of people from low-lying locations in England, the Netherlands, Belgium and France. The impacts in the coastal zone included severe cliff, beach and sand dune erosion, loss of some properties and disruption to transport links (page 129).

Seismic sea waves – tsunamis

A tsunami can be triggered when vast quantities of water are displaced. This can occur due to an earthquake affecting the sea bed or a large-scale underwater mass movement, such as the collapse of an underwater steep slope. The earthquake needs to be of a type that moves the seabed vertically, so not all underwater earthquakes result in a tsunami.

In deep water, tsunamis are hardly noticeable, having a wave height that is usually less than one metre and a very long wavelength, up to 200 kilometres. They travel at very high velocities, up to 800 km h^{-1}. Out at sea, they can pass beneath a vessel with little, if any, effect, but as they close on the shore their power is focused into a forward rush of water. *Tsunami* is a Japanese word for 'harbour wave'. Their height can climb to many metres and when the tsunami wave breaks, vast amounts of energy are released on to the coastline. The tsunami generated by the earthquake off the coast of Aceh province, Sumatra in December 2004 is estimated to have delivered about 1000 tonnes of water per metre of shoreline. The four to five metres of vertical movement of the seabed is estimated to have displaced some 30 km^3 of water, which then radiated out along the entire 1200 km length of the rupture in the rocks making up the seabed.

Waves in deep and shallow water

Waves in deep water

As air moves across a water surface, the friction between the air and water allows a transfer of energy from the atmosphere to the sea. This action creates pulses of energy which pass through the upper layer of water. Each wave thus possesses potential energy due to its height from wave crest to trough and kinetic energy caused by the movement of water within the wave.

When looking at waves out at sea beyond the near shore, it appears that, as waves pass through, they move water forward. Individual water particles, however, move in a circular motion at an equal velocity in all parts of the orbit (Figure 1.8). An object in the water basically bobs up and down as a wave passes.

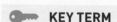

KEY TERM

Tsunami A wave created when a large mass of water is suddenly displaced.

▲ **Figure 1.7** The Great Wave off Kanagawa, woodblock print by Japanese artist Hokusai

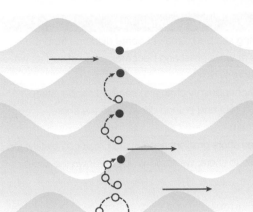

Water molecule in trough base. New wave front advancing.

Water molecule rising up wave front, but dragged slightly forward by wave advance.

Molecule 'sliding' down rear of wave, back into the next trough.

Molecule back in original position in orbit, as it reaches the trough base.

◀ **Figure 1.8** Circular motion in deep-water waves

The diameter of the circle travelled by a water particle decreases with increasing depth of water until a depth is reached at which the water is unaffected by the pulses of energy and conditions at the surface. This depth is known as **wave base**. Wave base is generally reckoned to be somewhere in the range of between one-half to one-quarter of the wave length. Wave base is important as above it waves are capable of eroding, transporting and depositing sediment, thereby carrying out considerable geomorphological work.

Waves in shallow water

As waves come close to the shore, they are modified as a result of decreasing water depth. The circular orbits of deep-water waves become more elliptical because of friction between the seabed and the water particles, with the forward element of the elliptical orbit being faster than the backward one. Wavelength and wave velocity both decrease. As energy has to be conserved, that is it can be neither created nor destroyed, it can only be transformed from one form to another. The wave's energy is transferred into an increase in wave height and consequently wave steepness. Eventually a situation is reached in which water is piled up to a height that over-steepens the leading part of the wave so that the orbits of the water particles are broken. The water rushes forward as the wave 'breaks' as **swash** followed by the **backwash**.

Types of breaking waves

Breaking waves are not all the same. The type of breaking wave depends on:

- wave steepness
- water depth
- the gradient of the shore.

Although various categories of waves exist, it is more realistic to think of wave type as existing on a continuum (Table 1.1; Figures 1.9, 1.10, 1.11).

 KEY TERMS

Wave base The depth at which water is unaffected by waves passing above.

Swash The forward movement of water once a wave begins to break.

Backwash The return flow of water away from the shore under the influence of gravity.

Breaker type	Description
Spilling	Steep; gently sloping sea bed; break at some distance from the shore; foam forms at wave crest and becomes a line of surf as wave approaches the shore.
Plunging	Steep; moderately or suddenly changing sea bed shape; steep-fronted; tend to curl over and plunge down on to the shore, producing lots of foam.
Surging	Gentle; steep-angled shore gradient; tend not to break completely; top of wave breaks close to shore; water slides up and down the shore.

▲ **Table 1.1** Types of breaker

▲ **Figure 1.9** Spilling breakers, Mawgan Porth, Cornwall

▲ **Figure 1.10** Plunging breakers, Beer, Devon

▲ **Figure 1.11** Surging breakers, Valentia Island, Ireland

Quite often the terms 'constructive' and 'destructive' are applied to wave type.

Constructive breakers	Destructive breakers
Lower wave height	Higher wave height
Long wavelength	Short wavelength
Low frequency: 6–8 per minute	High frequency: 12–14 per minute
Swash energy > backwash energy	Swash energy < backwash energy

▲ **Table 1.2** Characteristics of constructive and destructive waves

The idea behind the use of these terms is that constructive waves push sediment landward adding to the volume of beach sediment, while destructive waves have the opposite effect. This is a rather simplistic view of wave action and does not truly reflect what happens in the real world.

The relationship between process and form

The relationship between geomorphological processes, such as waves, and form, the shape or morphology of the landform such as a beach, is complex. Using terms such as constructive and destructive implies that the waves (the process) determine the shape of the beach (the form). However, one of the influences on wave type is the shape of the off-shore gradient, which includes the beach. Another influence on wave type is the particle size of the sediment making up the beach, for example the contrast between small-grained sand and pebbles. It is also the case that the same beach will receive different types of waves through time. In addition, the angle of wave approach can vary markedly through time.

The important significance of breaker type is the amount of energy that each brings into the coastal zone and how this impacts on the shore landforms. High-energy waves are able to carry out more 'work', such as sediment transport or erosion, than low-energy waves. In many ways, the use of terms such as high and low energy reflects more closely how waves and landforms relate to each other.

ANALYSIS AND INTERPRETATION

Study Figure 1.12, the relationship between wind speed and wave height.

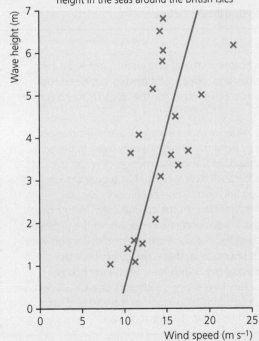

The relationship between wind speed and wave height in the seas around the British Isles

◄ **Figure 1.12** The relationship between wind speed and wave height

The data originate from the network of moored buoys, lightships and oil/gas platforms around the British Isles.

▶ Confidence values

Confidence limit	Critical value
0.05 level	0.388
0.01 level	0.549

(a) (i) Describe the relationship shown in Figure 1.12.

(ii) Following statistical analysis, a correlation coefficient of $r = +0.556$ was obtained. Interpret this result.

GUIDANCE

When interpreting a scatter graph, two aspects of the graph need to be considered: the direction of the relationship and its strength. Direction refers to whether the two variables are positively or negatively related. Figure 1.12 shows a clear positive relationship, with wave height increasing as wind speed also increases. The strength of the relationship is seen in the scatter of individual points. When the points are relatively closely grouped along the line of best fit, then the relationship is strong. The opposite is the case when the scatter of points is dispersed and especially if there are some clear **anomalies**. The scatter in Figure 1.12 indicates a strong relationship with no real anomalies.

The inclusion of the result from a statistical analysis allows a deeper interpretation to be conducted as it indicates how reliable the result is, that is whether the relationship is significantly different from what could have occurred by chance. The actual result is well above the critical value for the 0.05 confidence level and just above the value for the 0.01 confidence level. This means there is a less than 1 in 100 chance of the result having occurred purely randomly. It is possible, therefore, to suggest with quite a high degree of certainty that there is a strong positive correlation between wind speed and wave height.

(b) Explain how wave energy entering the coastal zone is affected by different factors.

GUIDANCE

It is important to appreciate that even if a statistical test indicates a strong correlation between two variables, this does not necessarily mean that there is a cause-and-effect relationship between the variables. It just indicates that a correlation exists and acts as a green light for the analysis to go further, confident that the relationship is worth investigating.

This question calls for an explanation of the factors that could influence wave energy. Clearly wind speed is important. Waves represent the effects of the transfer of energy from the atmosphere (winds) to the sea. The friction resulting from the moving air passing over water results in the orbital motion of water particles being initiated. The direction of the wind, its duration and the distance over which it has blown are key factors determining wave energy. However, as the scatter graph and correlation coefficient indicate, atmospheric energy is not the only factor influencing wave energy arriving in the coastal zone. Water depth and the offshore gradient are two important factors. As soon as a wave enters an area where the depth of water is less than the wave base, the circular orbits of water particles in a wave are affected by friction with the seabed. This reduces wave energy. The aspect of a stretch of coastline, that is the direction it faces, will influence the level of wave energy reaching it. On the west coast of the British Isles, west-facing coasts receive the full force of wave energy arriving from the Atlantic. Any bays facing south-east or east will receive much less wave energy. Most of their wave energy is the result of winds blowing from the east and in this direction the fetch is less and wind speeds are, on average, much less than the westerly winds.

🔑 KEY TERM

Anomaly A point that lies some way from the line of best fit. Anomalies can be termed exceptions or outliers.

Wave energy at the medium and local scales – wave refraction and diffraction

Wave refraction and diffraction are important factors because they affect the distribution of wave energy along a particular stretch of coast. This takes into account local factors such as the shape of the coastline.

Wave refraction

Whenever waves approach the shore at an oblique angle, the section of the wave in shallower water closer to the shore decelerates due to friction with the seabed. The remainder of the wave, in water deeper than the wave base, moves forward at a constant speed. As a result the wave bends or **refracts** so that its orientation is more parallel to the shape of the coastline. How well refracted waves become depends on factors such as the distance between where friction with the seabed begins and where the wave breaks.

The effect of energy inputs to the coastline can be seen when lines drawn at right angles to the wave crests are traced from offshore to the shore. These lines are known as **wave rays**.

▲ **Figure 1.13** Wave refraction along the Oregon coastline, north-west USA

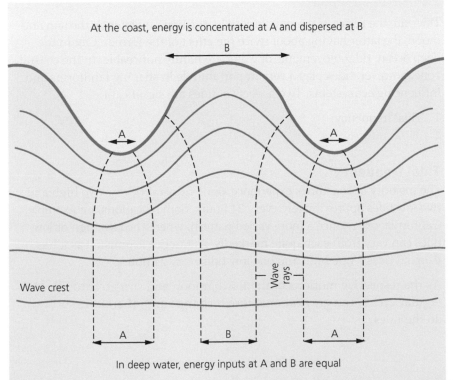

At the coast, energy is concentrated at A and dispersed at B

Wave crest

Wave rays

In deep water, energy inputs at A and B are equal

◀ **Figure 1.14** Wave refraction and energy patterns

🔑 **KEY TERMS**

Wave refraction The reorientation of wave crests as waves enter shallow water so that the wave approach becomes parallel to the shoreline.

Wave rays Lines drawn at right angles to wave crests; they are an indication of energy patterns. Sometimes called orthogonals.

Along relatively straight stretches of coastline, refraction is minimal, so energy distribution tends to be even along the shore. Waves approaching stretches of coastline consisting of bays and headlands are more subject to refraction and therefore variations in energy distribution. Energy is focused on the headlands where wave rays converge. Where wave rays diverge in bays, energy is more dispersed (Figure 1.14). Such variations in energy have significant implications for landform development.

Wave energy is also affected by the shape of the seabed in the coastal zone. The presence of deeper water over valleys or canyons allows waves to travel without losing their energy through friction with the seabed. Shallow areas offshore such as sandbanks mean that wave base is reached offshore and waves can start to break at some distance from the shore.

Wave diffraction

When a wave meets an obstacle, such as an island or a human-made offshore breakwater, **wave diffraction** occurs. Although the **lee** of the obstacle is protected from wave action, once the wave crest has passed the obstacle, energy is transferred into the sheltered area and wave action takes place. This impact of wave energy has implications for coastal management schemes that use offshore artificial reefs or breakwaters.

Tides

Tides are the result of the gravitational attraction on water by the sun and moon, the latter having about twice the effect of the former. Out in the deep ocean, tidal movements of water are hardly noticeable. In the coastal zone, however, tides play a very important role in shaping landforms and influencing ecosystems. Two aspects of tides are significant:

1 Tidal frequency
2 Tidal range.

Tidal frequency

The majority of coastlines experience **semi-diurnal** tides – two high and two low tides approximately every 24 hours. Some locations, for example California, experience a more varied pattern, where the two high or low tides can vary from each quite markedly. Antarctica is unusual as it has **diurnal** tides – one high and one low tide every 24 hours.

As the respective motions of the Earth, moon and sun go through regular cycles, the gravitational forces change and, therefore, so do the tides.

KEY TERMS

Wave diffraction Occurs when a wave passes an obstacle and its direction is altered so that wave energy is brought into the area behind the obstacle.

Lee The side of an object that is sheltered from a force such as wind or waves.

Tidal range The difference in water level between high and low water.

Semi-diurnal Refers to something having a twice-daily pattern.

Diurnal Refers to something having a daily or 24-hour cycle.

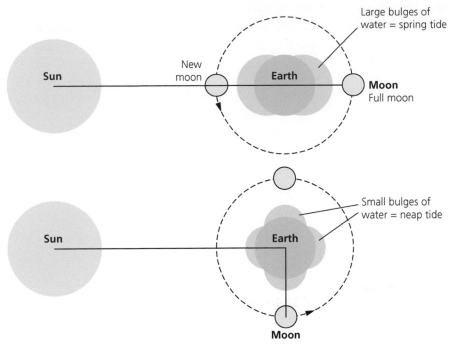

▲ **Figure 1.15** The formation of tides

Twice a month the sun and moon are aligned so that their gravitational forces are combined and act in the same direction. This produces **spring tides**, referring to the 'springing forth' of a tide. High tides are higher than average, low tides lower. Seven days later than a spring tide, the sun and moon are at right angles to each other with respect to the Earth. Their gravitational forces are therefore not acting together and so produce **neap tides**.

Tidal range

Tidal range is at its highest at the time of spring tides; neap tides produce the lowest range.

Tidal range varies considerably around the world (Figure 1.16). Enclosed seas such as the Mediterranean experience hardly any difference in sea level between high and low tides. On the other hand, the physical shape of the coastal zone including the underwater topography, can amplify tidal oscillation and generate tidal ranges over 10 metres. The greatest tidal range in the world occurs in the Bay of Fundy in north-east Canada, at around 16 metres. Around the UK, the Environment Agency maintains 44 tidal recording stations. The greatest tidal range in the UK occurs in the Severn Estuary, being between 14 and 15 metres.

 KEY TERMS

Spring tides These are nothing to do with the season but are the above average tides generated when the sun's and moon's gravitational forces act in the same direction.

Neap tides Twice monthly smaller tides when high tide is lower and low tide higher than average.

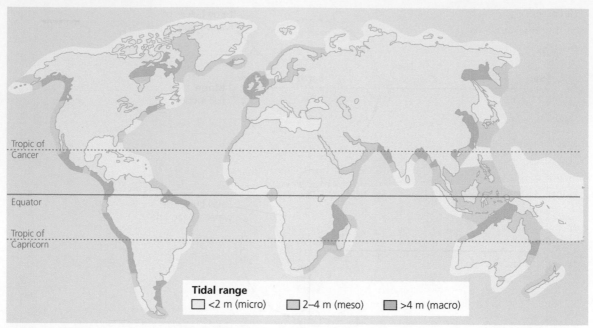

▲ **Figure 1.16** Global pattern of tidal ranges

Tidal range
☐ <2 m (micro)   ☐ 2–4 m (meso)   ■ >4 m (macro)

The significance of a coastline's tidal range is because of its effect on the extent of the intertidal zone and the time interval between tides. Both these factors influence weathering processes and biological activity. With a high range, more of the coastal zone is exposed to the impacts of alternate wetting and drying of rocks and the waves operate over a greater vertical extent bringing more energy into the coastal zone. Coasts with a high range tend to possess greater biodiversity in the intertidal zone as they provide more **ecological niches**. For example, a large tidal range is often associated with tidal flats and salt marshes.

Dynamic tide theory

As a tide is a wave of energy governed largely by interaction between the Earth and the moon, it might be expected that the crest of this wave lies directly under the moon as it orbits the Earth. However, this is not the case because of factors such as:

● ocean depth variations
● seabed topography
● shapes of the landmasses
● the ocean being broken up into deep basins (e.g. Atlantic, Pacific) separated by shallower shelves and continents.

The overall effect is that the tide is broken up into smaller-scale systems rather than one worldwide wave. As the gravitational pull of the moon passes across a basin, so does the tide. It is rather like holding a shallow dish of water and gently tipping it from side to side. The effect is that the water swirls around an **amphidromic point**. The greatest variations in tidal range occur furthest away from this point.

Because of the Earth's rotation, tides circulate around an amphidromic point: clockwise in the southern hemisphere; anti-clockwise in the northern hemisphere. Co-tidal lines join locations at which high tide occurs at the same time. Co-tidal lines radiate from an amphidromic point, usually being drawn at one-hour intervals starting at a particular point along a coast. High and low tide times therefore occur at different times along a coastline as the water swirls round the amphidromic point.

On the global scale there are six major amphidromic points, for example in the middle of the North Atlantic and halfway between South Africa and Antarctica. Smaller basins such as the North Sea also possess amphidromic points (Figure 1.17).

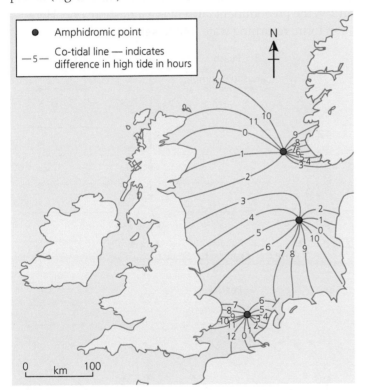

▲ **Figure 1.17** Amphidromic points and tidal patterns in the North Sea

Currents

There are several clearly identifiable flows of water in the coastal zone:

- tidal currents
- shore-normal currents
- longshore currents
- rip currents
- river currents.

Their water movements represent energy flows and so are important in the development of coastal landscapes and landforms.

In estuaries, the **flood tide** can pick up (entrain) sediment and carry it inland. Once high tide is reached, the current reverses and the **ebb tide** takes over. The velocities of tidal currents are relatively low at the start and end of each cycle and peak in the middle of the flood and ebb tides.

Where waves approach the shore with their crests parallel to the shape of the coastline, shore-normal currents exist. Water is carried up the beach but there has to be a return flow of water. At fairly evenly spaced locations along the shore, **rip currents** take water back out to sea after moving along the shore for a short distance. Their velocities can easily reach 1 metre per second, which is faster than an Olympic swimmer.

Where waves approach the shore at angles other than parallel to the shore, rarely greater than 10°, the predominant current is a longshore one. Here also, rip currents can form, returning water out to sea.

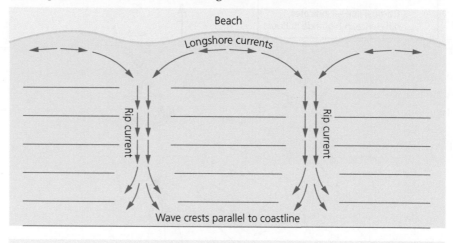

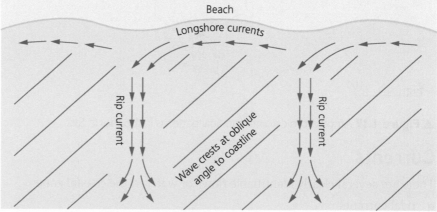

▲ **Figure 1.18** Rip current formation

In estuaries, river flows can be energetic, especially after heavy rainfall inland. As these currents pass into the coastal zone they interact with seawater movements, potentially causing very disturbed conditions.

3 Feedback, equilibrium and coasts

▶ *What happens to the coastal system when change occurs?*

Given that the energy inputs to the coast are so dynamic, change in the processes operating at, and landforms making up, the coast is inevitable. Such changes can be across the whole range of time scales, from the effect of a single wave breaking on a beach to the impact of tectonic adjustments that occur in geological time. Landforms develop a morphology (shape) which dissipates available energy. High energy leads to erosion and transport; low energy to deposition. As energy inputs change, landforms adjust in their shape. Some changes take place in just a few hours; others occur across thousands of years.

Feedback in the coastal system

Whatever the time scale involved, there is a very strong relationship between the energy coming into a location, the processes able to operate there and the landforms created. These process–form relationships involve feedback. Feedback takes place as a consequence of change in the system. Two types of feedback can be seen, positive and negative.

🔑 **KEY TERMS**

Positive feedback
Amplifies change.

Negative feedback
Restores a system to balance.

Positive and negative feedback examples

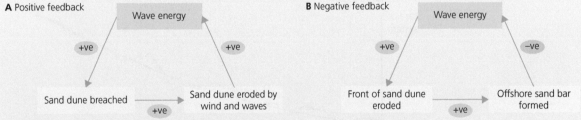

▲ **Figure 1.19** Feedback relationships between wave energy and sand dunes

Positive feedback can occur when wave energy increases, perhaps through a rising sea level. This leads to an increase in erosion of the dunes by removing the vegetation binding the sand. Wind action is thus able to erode the dunes further which makes the system more susceptible to wave action. The dune system therefore becomes more and more eroded.

Negative feedback can occur when wave energy increases. The increase in erosion of dunes by the waves releases sand which is then transported by the waves offshore. The sand is deposited as an offshore bar. As the incoming waves travel over the bar, the reduced depth of water means that more waves break before reaching the dune system. The reduction in wave energy allows the dune to recover as vegetation is able to establish and trap sand.

Equilibrium in the coastal system

The concept of **equilibrium** extends the use of the idea of coasts as open systems. It focuses on the balance between inputs and outputs and attempts to describe ways in which a system might respond to changes in that balance.

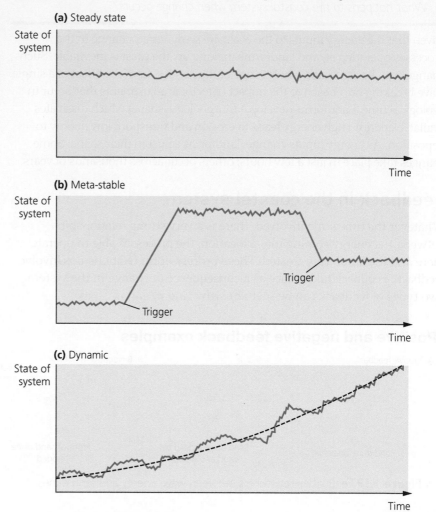

▲ **Figure 1.20** Types of equilibrium

The amount of energy entering the coastal system may be equal to energy dissipation without any change in the morphology or shape of landforms (Figure 1.20a). This **steady state equilibrium** continues until there is a change in the energy environment. Where and when sea cliffs receive more or less the same atmospheric and marine energy, the profile of the cliff tends to stay the same from year to year. A beach receiving similar wave energy from one year to the next undergoes seasonal adjustments, but its average annual gradient stays the same.

Sometimes a dramatic event occurs within the coastal zone that brings about substantial change to the coast. Sand and shingle can adjust rapidly to changing energy inputs. A high energy storm generating increases in wave height and periodicity can remove most of a beach in the space of a few hours. The result is a new shape, a wide flat beach. Subsequently the coastal zone adjusts to the new situation, a **meta-stable equilibrium** (Figure 1.20b). Wave energy is absorbed without any further net transfer of sediment.

Human activity can also bring about adjustment in the coastal zone. The construction of **groynes** over a six-month period triggers a change in the rate and volume of sediment moving along a coast. In turn this will influence beaches and cliffs by the groynes and further along the coast.

Time scales of energy change of a few hours will make no difference on solid rock coastlines. In the same way that rivers cannot adjust their channels when flowing over solid rock, so waves will make no impression on solid rock coastlines over short periods. Even if significant erosion occurs during a single storm, it will take thousands of years to achieve anything approaching an equilibrium form.

Even on hard rock coasts, equilibrium could in theory be achieved providing environmental stability lasted for long enough, perhaps over several thousands of years. Cliff recession would eventually form a shore platform wide enough to dissipate all wave energy before reaching the cliff line. Cliffs would then degrade through weathering and slope processes to achieve a new equilibrium form. But given the enormous climatic shifts of the past two million years as ice sheets advanced and retreated, and the accompanying changes in sea level, equilibrium on such a time scale is unlikely.

But change can be more gradual within the context of long-term changes. The coast could be in a state of basic or **dynamic equilibrium** (Figure 1.20c). Rising sea level is an example of this as wave energy reaches higher up the shore. Cliff and beach profiles will adjust as a consequence. Sediment input can increase as a result of deforestation in the catchment area of the river draining to the coast. This might cause an estuary to silt up more rapidly, with salt marsh growth accelerating.

So, part of the coast is in equilibrium, and part is not and probably never will be. The system has the freedom to adjust to energy changes on short time scales by shifting sediments (similar to a river flowing across its floodplain within a channel made up from alluvium), but little if any to bring about equilibrium on rocky coasts. As a result, in stormy conditions, the coastal system has surplus energy. Much of this surplus will be expended battering cliffs. The evidence for this is erosional features formed by cliff recession, such as caves, arches and stacks.

KEY TERMS

Meta-stable equilibrium Exists where a system changes dramatically between one state and another after the influence of a trigger and adjusts to that new state.

Groynes Barriers running across a beach down into the sea.

Dynamic equilibrium Involves change in a system but in a more gradual way than the dramatic change in meta-stable equilibrium.

We must not forget that wave energy also has an uneven spatial distribution on indented coastlines. Refraction concentrates wave energy on headlands but diffuses it in bays. Energy environments can change in just a few metres, adding further complexity to the coastal system. Thus variations in energy inputs are spatial as well as temporal.

④ Evaluating the issue

▶ *Assessing the relative importance of different energy sources in the coastal system.*

Identifying possible contexts, data sources and criteria for the assessment

This chapter's debate focuses on a core component of all physical systems: energy. The coastal system demonstrates, in various ways, just how significant energy inputs are to the way that it functions. There are several energy sources which need to be clearly described and explained, and a decision taken as to how to assess their relative importance.

- Possible contexts – in geographical assessment, it is important to consider scale. In the case of investigating energy inputs to the coastal system, a range of spatial scales are relevant, from the macro or global, through various meso scales (regional) e.g. the North Sea, to the micro or local such as a small bay, beach or reef. Scale can also be in terms of time, stretching from the macro or geological, through various shorter time periods such as the recent ice age, down to the time scale of a tidal sequence or an individual storm.

The greater range of spatial and temporal contexts will offer more opportunities to compare and contrast and thereby achieve an assessment of '…the relative importance of different energy sources'.

- Data sources – these will depend on the individual energy sources being considered. With a topic such as energy it is unlikely, if not impossible, to consider actual quantities of energy. However, there are some indicators of energy, such as wave height and wind speed.
- Criteria for assessment – as with data sources, quantifiable means of assessing relative importance are nigh on impossible to achieve. However, this does not mean that the issue is to be avoided – informed judgements should be made concerning such points as degree of erosion or deposition. Quantifiable evidence of possible outcomes of some energy inputs is available, such as rates of cliff recession.

Assessing the relative importance of solar energy

As for all natural systems, it is helpful to start with the key energy source of the sun. At the global scale insolation is not evenly distributed. The low latitudes receive more intense heating per unit of surface area than do the middle and particularly the high latitudes. Insolation in the polar regions is spread over substantially greater areas than at the equator. And for lengthy periods of time, virtually no energy from the sun enters these regions.

The significance of solar energy is that it drives many other natural systems. Variations in the heating of the Earth's surface at the global scale result in the global pattern of winds, with air moving from areas of high atmospheric pressure to areas with low pressure. This global circulation of air results in prevailing westerly winds affecting western Europe, western North America and western Chile. Equatorial locations such as much of the Indian subcontinent's coastline receive much less wind.

On a small scale, the differential heating of land and sea during the day sets up local onshore and offshore winds on a 24-hour cycle. Although not a major influence, such changes have some effect on the coastal system.

The sun also powers the hydrological cycle. Without its energy evaporation would not take place. In turn, with no moisture in the atmosphere, no precipitation would fall. The importance of this is in the context of weathering processes and the role of water in mass movement.

Nearly all ecosystems rely on solar energy as their starting point. Primary producers convert this energy via photosynthesis into structures which are then the food source for consumers. Solar energy is thus transferred through food chains and webs. The significance of living organisms is not just in the importance of biodiversity but in the impacts they have on landscapes and landforms. Coral reefs, mangroves, salt marshes and sand dunes would not exist without organisms.

Assessing the relative importance of waves

Waves represent a transfer of energy from the atmosphere to the sea. In this, it can be argued that they are subordinate to solar energy, which powers the atmospheric system. However, waves are a critical component in the coastal system because of the 'work' they carry out. It is just as important to recognise the significance of the absence of wave energy in a particular coastal location as it is to discuss the presence of wave energy in other places.

The role of wave energy in the production of landforms and landscapes must not be underestimated. Erosional process such as hydraulic action, abrasion and attrition rely on energetic waves reaching the shore. Sediment transport relies to a considerable extent on wave energy. Traction, saltation and suspension in water cannot take place without that water moving. In addition, without sediment erosion and transport there would not be deposition.

The contrast between high and low energy coastlines is essentially made in the context of wave energy. It is this input of energy, or the lack of it, which determines the ways the respective landforms and landscapes of these contrasting coastlines develop, for example a rocky, cliffed coast compared to an estuary. On a smaller scale, the contrast between the seaward and landward sides of a spit is essentially defined by wave energy.

Assessing the relative importance of tides

Tidal patterns and, in particular, tidal range, are significant because they influence the extent of the intertidal zone both vertically and horizontally. The more shore that is exposed between high and low tide, the more erosion, transport and deposition can take place. Tidal flows of water are very significant in estuaries, for example altering the mixing of salt and fresh water, which then influences ecosystems.

As with wave energy, tidal energy is related to the sun, though not in terms of insolation. Rather it is the gravitational influence of the sun that helps propel large-scale movements of water. However, the role of the moon is relatively more significant in tides than that of the sun. Although the sun is many times larger than the moon (diameter 400 times; volume 27 million times), its gravitational pull is less influential because it is so far away (400 times further away).

Assessing the relative importance of geothermal energy

Geothermal energy creates the rocks which are the foundation of a coastline. Rocks are created and destroyed in cycles. The rock cycle is a model that describes the formation, breakdown and reformation of a rock as a result of sedimentary, igneous and metamorphic processes. In addition, the role of tectonics has been and continues to be crucial. The opening and closing of oceans and land masses has affected the pattern of ocean currents. Present locations of land are made up of rocks formed under very different conditions to those being experienced today. The chalk cliffs of southern and eastern England relate to a time when these locations were much further south in the tropics. Many sandstones were formed in arid environments. Igneous and metamorphic rocks can be related to when the area they are found in was once an active plate boundary.

Tectonic energy has also brought about the folding and faulting of rocks. Without such massive releases of geothermal energy, many coastal landforms would not have developed as they have, such as various types of cliffs.

Over shorter time scales, such as the past few thousand years, tectonic forces have continued to exert a significant influence on the coastal zone.

- New land is created on the coast at locations such as Hawaii and Iceland as lava erupts and flows into the sea. Underwater earthquakes displace vast quantities of water, sending tsunamis racing across oceans. The impacts of the delivery of such enormous quantities of energy can completely remodel a coastline.
- Over the last ten thousand years or so, some coastlines around the UK have steadily risen upwards (at a rate of several mm per year), while others have sunk gradually. This is the result of isostatic rebound (see Chapter 5, pages 122–4), another process which depends on tectonic energy.

Arriving at an evidenced conclusion

The highly dynamic nature of the coastal system is down to the substantive transfers of energy that take place within its confines. Landforms and landscapes that develop along coastlines are so diverse because of energy variations over long, medium and short time scales.

There is no coastline without geothermal energy. Rocks need to be formed and lifted above the surface of the sea. No cliffs can exist without the rock cycle, nor any beaches. The very existence of the current pattern of land and sea is due to tectonics.

However, without insolation, neither the atmospheric system nor the hydrological cycle could function. Winds, waves, rainfall, water flowing off the land all rely on solar energy. In addition the biosphere relies on insolation to provide the energy that can then flow throughout ecosystems.

Pearson Edexcel

AQA

OCR

WJEC/Eduqas

A case can be made for solar energy to play the most significant role, through the agencies of winds and waves in particular. However, take away or alter any one component and feedback starts to operate, bringing about changes. It is this variation in relative energy within the coastal system that makes it so dynamic. No two waves are exactly the same, and the structures and lithologies of apparently identical rocks can vary in subtle ways that lead to slightly different landforms developing.

Finally, over the course of the past two hundred years, human activities have been intervening in the coastal system either deliberately or unintentionally, to add yet another source of 'energy' into the coastal system. Along increasingly densely populated coastlines, it is human activities that are perhaps the most significant factor. Perhaps the most dramatic influence is only just beginning, that of the rise in sea level associated with anthropogenic global warming.

 KEY TERM

Isostatic rebound The upward movement of land as weight is removed, e.g. through the melting of ice or weathering and erosion of rocks.

Chapter summary

✔ Coasts function as open systems. Flows of energy and materials enter and leave the system; as these flows interact, distinctive coastal landforms and landscapes are produced.

✔ The time scales over which processes operate vary from millions of years to just a few seconds. A systems approach helps understand how such variation in time scale shapes landforms and hence landscapes.

✔ Waves generated by winds are a major energy input into the coastal system. As waves come close to shore, they break and their energy enters the coastal zone. The energy they bring varies both spatially and through time.

✔ Tides and currents move both water and sediment and vary around the world; tidal range is a significant component in a location's coastal system, influencing the distribution of processes across the coastal zone.

✔ The amount of energy coming into any part of the coastal zone influences the processes operating and the landforms and landscapes formed. Two types of feedback can occur, positive and negative, each being a response to something changing in the system. Equilibrium in the coastal system depends on the balance of energy within the system; stability or disturbance can occur.

Refresher questions

1   Identify two of each of the following in the coastal system: inputs; stores; processes; outputs.

2   Outline how wave energy along a coastline is affected by the length of fetch.

3   Describe what happens to the pattern of water particle movement when a wave approaches and then reaches the shore.

4   Explain the significance of swash and backwash when considering the impacts of waves on a coastline.

5   Explain how wave refraction influences the distribution of wave energy along a coastline.

6   Outline the importance of tidal range to coastal processes.

7   Describe and explain the difference between positive and negative feedback.

8   Explain how dynamic equilibrium might operate within the coastal system.

Discussion activities

1   Discuss how energy inputs from past events can continue to influence the contemporary coastal system. You could focus on a stretch of coastline known to you. Research its geology and then draw a timeline covering the periods when the rocks of the coast were formed. Identify key events/periods in the past when there were significant energy inputs.

2   Consider the advantages and disadvantages of using a systems approach to understanding coastal landforms and landscapes. Focus on the value of looking at flows of energy and materials through a system. Then think about how human beings interact with the coastal environment and how such interactions can be included in a systems approach.

3   In small groups, discuss variations in wave energy due to aspect and fetch that coastlines around the British Isles are likely to receive. Use atlas maps and images from Google Earth as the basis of your discussion.

4   Discuss what elements in the coastal system are likely to respond either rapidly or only slowly to changes in energy inputs.

5   Discuss ways in which wave energy influences informal representations of places. For example, research pictures painted by artists that represent different 'moods' of the sea, such as a storm. Additionally, find pieces of music or excerpts of fictional text that portray contrasting levels of wave energy.

FIELDWORK FOCUS

Investigating the ways that elements in the coastal system respond to changes in energy and/or material inputs is not straightforward. Possible opportunities are suggested in subsequent chapters, for example in the context of beaches. It is possible to look at changes in the outline of a stretch of coastline as evidence of how feedback processes might be operating. Comparing maps over a time period of 100 or more years can reveal significant adjustments in the coast. Ordnance Survey maps date back to the early nineteenth century and can be supplemented by maps drawn in earlier times. When comparing maps across such a range of time, it is important to notice any differences in scales so that you are comparing like with like as regards distances for example. For some locations, coastal change might also be investigated by analysing old postcard images. These secondary sources could be compared to the present-day context, which would be your primary data source.

Further reading

British Geological Survey – various pages on the site offer much detailed information about geology: see www.bgs.ac.uk The marine science area has some dedicated pages offering information about seabed geology.

Channel Coast Observatory : https://www.channelcoast.org

Coastal Wiki: http://www.coastalwiki.org/wiki

Luijendijk, S.A., Hagenaars, G., Ranasinghe, R., Baart, F., Donchyts, G., Aarninkhof, S. (2018) *The State of the World's Beaches*, Scientific Reports, 8, 11 [open access]

Masselink, G.L., Hughes, M.G., Knight, J. (2011) *An Introduction to Coastal Processes and Geomorphology* (3rd edition). Abingdon: Routledge

Open University (1999) *Waves, Tides and Shallow Water Processes* (2nd edition). Milton Keynes: Open University

Pilkey, O.H., Neal, W.J., Cooper, J.A.G., Kelley, J.T. (2011) *The World's Beaches: A Global Guide to the Science of the Shoreline*. Berkeley and Los Angeles: University of California Press

CHAPTER 2

Coastal system dynamics – processes at work in the coastal zone

Large amounts of energy and materials flow through coastal environments, with the effect that changes in the morphology of landforms are common. The interaction of marine erosion, sub-aerial processes and mass movements creates some of the most dramatic slopes in the world. Sediments are regularly and frequently transported and deposited by both water and air movement, contributing to the dynamic nature of the coastal zone. This chapter:

- examines marine erosional processes
- investigates sub-aerial weathering and mass movements
- analyses different sources and types of coastal sediment
- explores processes of coastal transport and deposition
- evaluates the extent to which wave energy is the most significant influence on sediment transport in the coastal zone.

KEY CONCEPTS

Systems Groups of related components. In physical geography, systems tend to be 'open', that is having both inputs and outputs in the form of energy and materials. A sediment cell is such a system with inputs (e.g. wave and wind energy), stores (e.g. sediment on a beach) and outputs (e.g. sediment transported out of the cell).

Equilibrium The state of balance within a system. If a sediment cell loses sediment, for example through human activities extracting sand and shingle, the stores and flows of sediment within it are disturbed. This can have significant impacts for the rate of cliff erosion, for example when a beach loses material this allows more wave energy to reach the cliff.

Feedback An automatic internal response to change in a system. For example, undercutting of a marine cliff by waves leads to a steepening of the slope angle of the cliff. As undercutting leads to a reduction in stability, the increased stress on the slope system potentially exceeds the threshold for slope failure. Positive feedback results in change to slope morphology.

Threshold A critical 'tipping point' in a system. If a cliff receives a substantial input of water, such as from a period of intense rainfall, the additional weight of that water can 'tip' the slope past its equilibrium point (stability threshold). Shear stress is then greater than shear strength, and mass movement occurs.

① Marine erosional processes

▶ *What erosional processes operate in the coastal zone?*

Energy inputs into the coastal zone are a very important factor influencing which processes are significant in a location. The ability and extent to which waves can **erode** depend on three variables:

- wave energy
- geology of the coastline
- morphology (shape) of the coastline.

The 'significant wave height' measured at a coastal location indicates how much force moving water might exert on the coast (page 6). However, it is important to appreciate that for most of the time, erosion is not occurring. It is now recognised that many landscapes are relatively 'quiet' and it is the infrequent and irregular very high energy events that carry out the 'work' required to bring about change to landforms such as cliffs. Even in those coastal locations where the rocks are relatively weak (e.g. clay), erosion is not constant (page 180).

The geological variable includes the **lithology** and **structure** of rocks.

Coastal morphology refers to features such as whether there are pronounced bays and headlands. It also includes the immediate offshore underwater features such as ridges and valleys.

KEY TERMS

Erosion The wearing away and/or removal of rock and other material by a moving force such as water.

Lithology The study of the physical and chemical composition of a rock as well as its texture.

Structure Features such as bedding planes, joint pattern, folds, faults and angle of dip of the rock.

Hydraulic action Involves the force of moving water weakening and dislodging rock.

Processes of erosion

Hydraulic action

The process of **hydraulic action** involves the movement of water, without the involvement of rock particles. The key factor here is purely the force of moving water. Sometimes referred to as wave pounding, the alternating application and release of water pressure weakens rock. In extreme conditions the pressure can reach levels of around 11 000 kg/m². The impact of a mass of water can dislodge fractured and loose rock fragments, termed wave quarrying.

▲ **Figure 2.1** High energy waves breaking on a cliff

Added to the force of water are **pneumatic** forces. Air can be trapped and compressed between forward-moving water and a cliff face. This is a particularly effective force when the cliff is made up of well-jointed rock. As the wave retreats, the pressure is suddenly released, which weakens the cliff face.

If the waves are extremely large, **cavitation** occurs. Bubbles, at great pressure within the wave, collapse. This generates shock waves that erode rock surfaces with similar effects to hammer blows.

Abrasion

Breaking waves pick up and carry sediment such as sand, gravel and pebbles. As the moving water drags the sediment over rock and as sediment is hurled at a rock face, a scouring action takes place, known as **abrasion** (also called **corrosion**). How effective abrasion is depends on wave energy and on the availability of sediment. Larger calibre sediment (boulders and pebbles) is moved only when water has plenty of energy, for example in an intense storm. Smaller sediment (sand and gravel) is moved more frequently and regularly by lower energy waves.

Attrition

Individual sediment particles collide with each other as they are moved around by water. When this happens fragments are broken off, which reduces the size of the particles. It tends to be the smaller particles that are carried in water as suspended sediment. On a beach, particles can be rolled up and down as swash and backwash occur. This process, called **attrition**, produces smoothed and rounded sediment. Sediment is also broken and rounded when it is involved in abrasion.

Very large boulders scattered along several coastlines have been researched in order to identify their origin. It would seem that these boulders, one example in the Bahamas weighing as much as 1000 tonnes and measuring 13 × 11.5 × 6.5 metres, were moved by either a tsunami or an incredibly intense storm in the recent geological past around 125 000 years Before Present day (BP). Recent research confirms that large boulders weighing several hundred tonnes were moved tens of metres during the winter storms of 2013–14 on the Aran Islands, north-west Ireland.

Bioerosion

Bioerosion is caused by the activity of biological organisms living in the coastal zone.

- Gastropods (e.g. whelks) and echinoids (e.g. sea urchins) rasp a rock surface as they graze algae growing on the rock. This gradually removes thin layers of rock. Estimates of between 0.4 to 2.0 mm/year have been suggested for the lowering of limestone rocks in the Balearic Islands, Spain due to grazing.

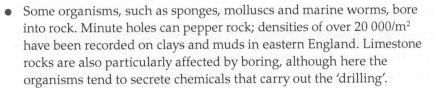

- Some organisms, such as sponges, molluscs and marine worms, bore into rock. Minute holes can pepper rock; densities of over 20 000/m^2 have been recorded on clays and muds in eastern England. Limestone rocks are also particularly affected by boring, although here the organisms tend to secrete chemicals that carry out the 'drilling'.
- The mechanical action of plant roots forcing their way into crevices and cracks can break open rocks. This allows other destructive processes to become effective, in particular due to water accessing the rock and a larger surface area being open to attack.
- Seaweeds can anchor themselves on to rocks. At times of high wave energy, the movement of the seaweed can prise off the piece of rock the plant is attached to.
- Some species of birds and mammals, for example puffins and rabbits, burrow into cliff faces, making mechanical collapse more likely as well as allowing chemical processes to act on material along the burrows.

 # Sub-aerial weathering and mass movement

▶ *What sub-aerial weathering processes and mass movements operate in the coastal zone?*

The climate and weather of the coastal zone are important inputs to the coastal system. In regions of low wave energy and tidal range (pages 6 and 16), such as the Mediterranean and south-east Africa, **sub-aerial weathering** can play a dominant role in landform development. The same range of weathering processes that occur inland may be present at the coast. However, the presence of seawater and the effects of tides in wetting and drying rock bring additional destructive influences. There is also the effect of past climates, such as the legacy resulting from the advance and retreat of ice sheets.

Sub-aerial weathering

A range of weathering processes can be active on a stretch of coastline. However, the type of rock making up the coast, together with the climate, influences the processes that occur there.

For convenience, weathering is often divided into two main categories, although the two are closely linked and often occur simultaneously:

- Physical/mechanical – rock is broken down by physical force.
- Chemical – a chemical reaction occurs, changing minerals into different forms.

 KEY TERM

Sub-aerial weathering
Processes occurring at the land surface. Weathering processes are those that mechanically disintegrate and/or chemically decompose rocks where they are (in situ).

▼ **Table 2.1** Effects of physical/mechanical weathering processes on coasts

Weathering process	Effect
Crystal growth	Expansion of crystals (e.g. salt) when seawater collects in cracks in the cliff face → solution evaporates → crystals form. Crystal growth → pressure exerted on the rock. Most effective on porous rocks, e.g. sandstones, and locations where temperatures promote evaporation. Probably the most significant weathering process in the coastal zone.
Freeze-thaw	Repeated freezing and thawing of water; most effective on high-latitude coasts experiencing high precipitation.
Wetting and drying	Expansion and contraction of minerals, often due to rising and falling tides; most effective on clay in macro-tidal locations.

▼ **Table 2.2** Effects of chemical weathering processes on coasts

Weathering process	Effect
Solution (also known as corrosion)	The dissolving of minerals, changing their state from solid to liquid. Solubility depends on temperature and acidity of water. Limestones such as chalk are particularly affected by solution (process of carbonation), although less so in seawater.
Hydration	Absorption of water by minerals weakens their crystal structure, making rock swell and more susceptible to other weathering processes where weaknesses occur e.g. in shales and clays.
Hydrolysis	A reaction between mineral and water related to hydrogen ion concentration in the water; particularly affects the feldspar minerals in granite, producing clays which are more susceptible to further weathering and erosion.
Oxidation/reduction	Adding/removing oxygen. Oxidation results from oxygen dissolved in water, particularly affecting rocks with high iron content; reduction is common under waterlogged conditions.
Chelation	Organic acids produced by plant roots and decaying organic matter bind to metal ions, causing the rock to decompose.

▲ **Figure 2.2** Limestone weathered by a combination of physical, chemical and biological processes, Dorset

Biological activity often increases rates of weathering. The production of organic acids directly causes chelation while erosion, such as burrowing, allows water greater access into rocks.

Wetting and drying

The tidal cycle leads to alternate wetting and drying of rocks exposed between high and low tides. This cycle has a significant effect on the development of landforms in the coastal system. The zone of the coast affected extends inland from high tide to where spray is thrown. As a result, tidal range and meteorological factors, such as air temperature, are important. For example, they dictate how quickly evaporation takes place.

Mass movement

The coastal system has some of the most dramatic slopes in the world in the form of towering cliffs. There are also gentle slopes at the coast. The full range of slopes in the coastal zone is subject to the downslope movement of material under the influence of gravity, known as **mass movement**. This does not include material moved directly by an agent such as a river, glacier or sea wave.

In mass movement, a quantity of rock or soil moves as a single unit, although there might be some movement of particles or fragments within it. Whether material on a slope moves or remains where it is, is determined by a balance between shear strength and shear stress. The ratio of these forces is known as the safety factor:

$$\text{Safety factor (SF)} = \frac{\text{sum of forces resisting movement}}{\text{sum of forces driving movement}}$$

If SF > 1 then movement will not occur; if SF ≤ 1 then movement will begin. The main forces influencing movement are:

- gravity
- slope angle
- water content of the slope material.

On any slope the weight of the material tends to pull it downslope. Downslope movement is proportional to the weight of the block and the slope angle. The heavier the material and steeper the slope, the more likely it is that mass movement will occur. Water adds weight to slope material (a litre of water weighs one kilogram). Additionally, water acts as a lubricant which reduces frictional resistance within the slope material. In material such as clay, the spaces between its very small particles can become filled with water, which forces them apart, something known as pore pressure. Increasing pore pressure makes mass movement more likely.

Mass movements occur at a range of rates from very fast to the almost imperceptible (Figure 2.3).

KEY TERMS

Mass movement The downslope transport of material under gravity.

Shear strength The internal resistance of material to movement.

Shear stress The forces attempting to move the material downslope.

Pearson Edexcel   AQA   OCR   WJEC/Eduqas

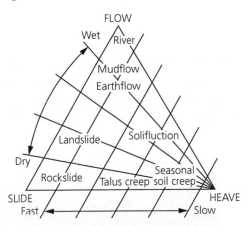

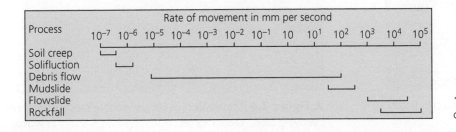

◄ **Figure 2.3** A classification of mass movements

Several variables interact to influence the type of mass movement that occurs at any one location:

- calibre (size) of the slope sediment/material
- water content of the slope materials
- height of the slope
- slope angle
- vegetation cover.

Types of mass movements include:

- Rock fall – steep, unvegetated rock faces become weakened by weathering; blocks eventually dislodge and crash down.
- Rock slide – blocks or sheets of rock slide down the cliff face along seaward-dipping bedding planes.
- Rock toppling – blocks or columns of rock, weakened by weathering, fall seawards.
- Rotational slides and slumps – sections of a cliff give way along a well-defined concave slip surface. The fallen material stays as an identifiable mass with a back-tilted flat top until further weathering and erosion act on it. Slumps occur when a section of cliff collapses as a jumbled mass of material. Rotational slides and slumps are common where **permeable rock** lies over **impermeable rock** or where the slope consists of unconsolidated material such as glacial deposits.

▲ **Figure 2.4** Rock fall, East Devon

KEY TERMS

Permeable rock Rock through which water can flow.

Impermeable rock Rock through which water cannot flow.

Regolith The layer of loose material, including soil, above bedrock.

- Flows – unlike slides and slumps, flows tend not to be broken up into sections; commonly occur in saturated clays and unconsolidated sediments.
- Creep – extremely slow downslope movement of regolith.
- Solifluction – slow downslope movement of waterlogged **regolith**. It is common in the summer in high latitude areas where the top layer of material has been frozen over the winter. It is also common when an ice age is ending and along some mid-latitude coastlines it exists as a relict deposit.

▲ **Figure 2.5** Rock slide-affected cliff, South Wales

▲ **Figure 2.6** Rotational slides in permeable sandy sediments overlying clays, West Dorset

ANALYSIS AND INTERPRETATION

Study Figure 2.7 which shows average rates of cliff retreat per year by rock type.

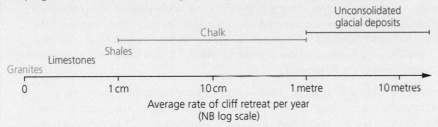

▲ **Figure 2.7** Average rates of cliff retreat per year by rock type

(a) Using Figure 2.7, compare the average rates of cliff retreat for granites and unconsolidated glacial deposits.

GUIDANCE

The diagram looks a straightforward one to interpret at first sight and in some ways it is. There are only a handful of different rock types to consider and just one variable for each type, average rate of cliff retreat. However, as with all such analyses, it pays to spend a little time making sure you understand the data. In this figure, the key point to note is that the rates of cliff retreat are plotted using a log scale. The scale does not increase arithmetically, that is 1, 2, 3 and so on. The rates of retreat increase by a factor of ten.

If there is the requirement to 'compare' in the question, then it is important to compare in your answer. Do not write separate descriptions of the rates of cliff retreat. For these data, the comparison between granite cliff retreat and that occurring in unconsolidated glacial deposits is stark. Granite cliffs retreat at a much slower rate, by just a few millimetres per year, whereas cliffs made up of unconsolidated glacial material tend to retreat much more rapidly, between 1 and 10 metres per year.

(b) Explain why chalk cliffs have a relatively wide range in their rates of cliff retreat.

GUIDANCE

Figure 2.7 indicates that chalk cliffs retreat at rates of between 1 and 100 centimetres per year. This requires you to think of possible reasons why some chalk cliffs are relatively stable while others undergo relatively rapid recession. Rocks vary in their lithology and structure, both between one rock type and another, and within the same rock type. No rock is the same wherever it is found. The conditions under which a rock is formed will be broadly similar whenever and wherever that rock type exists, but not exactly the same. Slight differences in mineral composition (lithology) and density of joints/bedding planes (structure) will occur. Such contrasts can influence the effectiveness of sub-aerial weathering, in the case of chalk solution/carbonation weathering for example. Another factor explaining why some chalk cliffs retreat more rapidly than others could be contrasts in marine erosion. If a chalk cliff is protected at its base by a wide beach, then wave energy is likely to be mostly used up before the incoming rush of water is able to reach the cliff. Other chalk cliffs will be subject to much more active wave action and so will be undercut and suffer from collapse. Chalk cliffs on an exposed headland will retreat more rapidly than those in a relatively sheltered location.

(c)   Explain the value of comparing average rates as in Figure 2.7.

GUIDANCE

The average, otherwise known as the mean, gives a simple overview of the whole data set. It is useful in the context of the rates of cliff recession as it allows direct comparison among the five rock types. However, the size of the data sets used to calculate the averages is not known. It may be that for one rock type, only a few figures of cliff retreat were available. In this case, a couple of exceptionally high or low values would make the average figure not very representative of the whole group.

Average rates of cliff retreat are helpful in giving an overview as to how resistant a particular geology is over a period of time. However, it is important to consider the point that for much of the time cliffs, like other coastal landforms, are not that active. Change to landforms, such as cliff collapse, tends to happen as occasional high-energy events. After such an event, the landform tends to revert to a relatively 'quiet' state of equilibrium.

 Sources and types of coastal sediments

▶ *Where does coastal sediment come from and what types of sediment are there?*

Sediment produced by erosion, weathering and mass movement is an important component of the coastal system. Because sediment is unconsolidated, that is it is loosely arranged, it can be transported by waves, currents and the wind. Mass movements also move sediments down slopes.

 **KEY TERM**

Denudation The wearing away of the Earth's surface by weathering, erosion, mass movement and transport of material.

Sources of coastal sediments

It might seem as if the obvious sources of coastal sediments are coastal landforms such as cliffs. However, about 90 per cent of coastal sediment originates from the **denudation** of inland areas. The broken down material is transported into the coastal zone by rivers (Figure 2.8).

At locations where wave energy is low, especially in the tropics, river sediments are particularly significant. Rivers such as the Niger, Ganges and Amazon carry very high suspended loads into the coastal zone.

There are, however, stretches of coastline where local geology plays an important role in supplying sediment. Current volcanic activity along stretches of the coast has led to the breakdown of lava to form black beaches such as at Panalu'u, Hawaii and Reynisfjara beach, south-east Iceland (Figure 2.9). Along parts of the southern Californian coast, recent research has shown that as much as 50 per cent of beach sand has come from the collapse of cliffs composed of relatively weak sedimentary rocks. Beaches and dune systems can erode and their sediments move from the land to the sea. Some sediments, for example shell, are generated entirely within the coastal zone.

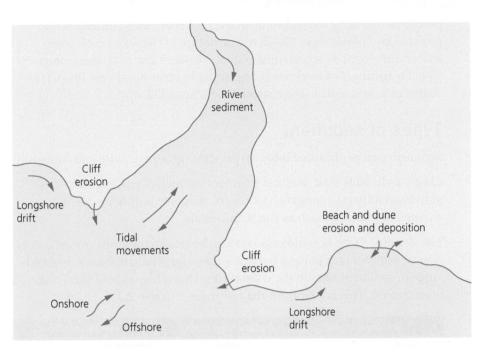

▲ **Figure 2.8** Sources of coastal sediments

◀ **Figure 2.9** Black basaltic sediment, Reynisfjara beach, south-east Iceland. This sediment has its origins in the basalt erupted from the mid-oceanic ridge in the North Atlantic, on which Iceland sits. Basalt contains minerals that have dark colours, such as olivine and hornblende

Transport of sediment from offshore towards the land can also take place. Such sediments might be former beach or river material that initially had been carried offshore. Exceptionally energetic waves can bring sediment onshore but most onshore movement is due to the rise in sea level following the melting of the ice sheets and glaciers of the Quaternary period. Long-term dynamism in the water cycle meant that water moved from its frozen store on land into the seas. This is an important reminder that landforms existing today cannot always be understood by considering only the processes operating today. Events in the past continue to influence the present-day coastal zone. The dynamic nature of the water cycle over shorter time periods is becoming only too evident due to contemporary global warming. Sea level rise is beginning to have significant impacts on sediment movements in the coastal zone (pages 132–4).

Types of sediment

Sediment can be classified into two principal groups, clastic and biogenic.

Clastic sediments exist as either fragments of rock of varying size or as individual mineral grains such as quartz. Biogenic sediments are the remains of materials such as shells and corals.

The size and shape of sediments can also be used to describe material. Size is most commonly taken as a particle's diameter. There is such a very wide range of sediment sizes in the coastal zone that a logarithmic scale has been created. This accounts for the categories in Table 2.3.

Particle name	Actual size (mm)	Relative size
Boulders	2048	very large
	1024	large
	512	medium
	256	small
Cobbles	128	large
	64	small
Pebbles/shingle	32	coarse
	16	medium
	8	medium
	4	fine
Sand	2	very coarse
	1	coarse
	0.5	medium
	0.25	medium
	0.125	fine
	0.063	very fine
Silt	0.031	very coarse
	0.016	coarse
	0.008	medium
	0.004	fine
Clay	0.002	very fine

▲ **Table 2.3** Particle names and sizes

When studying sediment that is pebble-sized or larger, the three axes of an individual particle are used to help describe shape (Figure 2.10).

These three measurements can be used to distinguish the shape of a particle. The four categories of shape are:

- Rod – long and thin
- Sphere – ball-like
- Blade – long and flat
- Disc – round and flat.

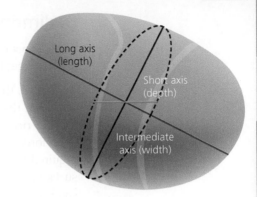

▲ **Figure 2.10** The axes of a particle

▲ **Figure 2.11** A mix of pebble shapes, Beer beach, East Devon

 # Transport and depositional processes in the coastal zone

▶ *How is sediment moved and deposited in the coastal zone?*

Sediment movement occurs when there is sufficient energy to **entrain** particles. Moving air or water can apply sufficient force on a particle to lift or drag it from where it is resting. The velocity required to entrain a particle depends on the size of that particle.

Once on the move, less energy is needed to keep the particle moving. However, a drop in velocity below that required for movement results in deposition of the particle.

 KEY TERM

Entrainment The process by which particles are picked up from a stationary position and moved, by a flow of air or water.

Sediment movement

In the coastal zone, moving water often transports sediment. Winds and ice are also agents of transport, for example along some high latitude coasts.

In general, the larger the particle, the more energy is required to entrain it. However, the smallest particles, silt and clay, are exceptions. These particles are electrically bonded together, which makes them more cohesive. Therefore, more energy is required to lift them off the seabed than is required to lift sand grains. Silt and clay particles also present a smoother surface for the water to push against compared to sand grains. Once these smaller particles are moving, relatively little energy is required to keep them moving. Larger sediment, such as pebbles and cobbles, is soon deposited once water velocity slows.

Breaking waves can cause much turbulence, which makes sediment transport in the coastal zone a complex process. Water movement is not just in one direction. Swash and backwash carry sediment up and down a beach. Water can move along a shore and flow back out to sea, via rip currents, carrying sediment as it goes.

Types of movement

Sediment movement essentially falls into one of four types: traction, saltation, suspension and solution.

Traction is when larger calibre sediment tends to roll and slide along the seabed, but only intermittently, as and when high energy conditions exist. It is possible to hear pebbles being moved on a shingle beach when high energy waves break. When the same beach receives low energy waves, few pebbles are moved.

Saltation results in a skipping motion of sand grains along the seabed, but or on a dry beach.

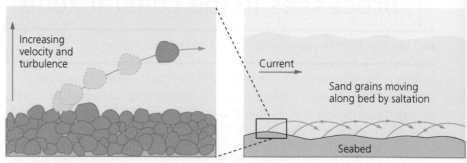

▲ **Figure 2.12** Saltation and the transport of sand grains

Individual sand grains move along the surface in an arc-shaped trajectory. When an individual grain lands, it disturbs more grains, causing these to be transported. Relatively short distances are travelled in any one 'skip', however the process is cumulative and vast quantities of sand can be moved. Saltation is a very significant aeolian process and important in understanding sand dune evolution (page 96).

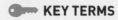

KEY TERMS

Traction The movement of larger calibre sediment (for example shingle and pebbles).

Saltation When sand grains are picked up, moved forward and then dropped down again, either by wind or water.

Suspension Occurs when sediment, usually no larger than sand grains, is kept moving by wind or water.

Solution Occurs when minerals are dissolved and then transported by water.

Aeolian Refers to the action of the wind, such as sediment being wind-borne or wind-deposited.

Sediment, usually sand grains and finer particles, is kept moving by turbulent water. The finer clays and silts continue to move in all but very low energy conditions. When high wind speeds occur over a dry beach, sand grains can be suspended but it is difficult to distinguish this from saltation transport.

When minerals are dissolved in seawater, the resulting solute is transported by the sea. Limestones in particular are susceptible to the chemical processes that generate solution.

On-shore movement of sediment

Below wave base (page 9), sediment on the seabed is not moved by passing waves. There are, however, currents of water that can transport material from accumulations offshore towards the shore as part of a sediment cell.

Perhaps the most significant on-shore movement of sediment relates to the last ice age. When the final major ice advance was at its maximum, some 18 000 years BP, sea level was between 100 and 120 metres below what it is today. The very extensive areas of dry land exposed as a result had a covering of weathered and broken up material. As sea level gradually rose, this material was picked up and carried on-shore. The sediment was altered in shape and size by attrition and, when sea level stabilised, deposited. There are landforms built from significant accumulations of such 'relict' sediment in the mid-latitudes, the presence of which can only be explained by these processes. Chesil Beach in southern England is one such landform. This is another example of the importance of past processes on present-day landforms and landscapes.

Offshore movement of sediment

There are three ways by which sediment tends to move offshore:

- During storm conditions, high energy waves are capable of transporting much material out to sea. A severe storm coinciding with a high tide can not only lower a beach but can also remove other unconsolidated material, such as sand dunes, from just inland.
- Sediment moved along a coastline might reach a river estuary. The river current might carry the marine sediment, as well as its own load, offshore to be deposited in deeper water. The energy of the river flow will determine how effective a process this is. The Amazon is estimated to carry 3 to 3.5 million tonnes of fine sediment every day into the Atlantic Ocean. Most of this is transported many kilometres beyond the coastal zone, but some 20 per cent is moved westwards by the Guyana Current along the north coast of South America. This has created the world's longest mud shoreline, of some 1600 km.
- The seabed is not uniformly flat. A relatively common feature is a submarine canyon extending offshore. Water flows down and along the canyon, carrying with it sediment to deeper offshore locations such that it is lost from the coastal system. The coastal zone of southern California has many such canyons extending out into the Pacific along which large volumes of sediment move.

 KEY TERM

Load The material transported by an agent such as a river, ice, the wind or the sea.

Longshore movement of sediment

When the angle of wave approach to the shore is oblique, the advancing swash moves up the beach at the same angle. Sediment is entrained by the moving water and transported up the beach in the swash. The wave eventually runs out of energy because of friction and the gradient of the beach profile. Some sediment is likely to be deposited at the point where the swash finally ceases. Smaller particles might continue to move with the water that returns to the sea as backwash. This flow returns to the sea directly under the influence of gravity, which acts perpendicular to the beach slope. The next wave then breaks and the process is repeated. Sediment is once again entrained and transported along the shore in the direction of wave approach.

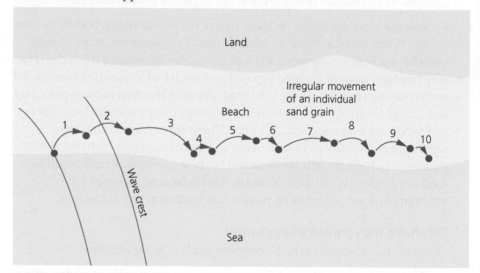

▲ **Figure 2.13** Longshore drift

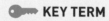
 KEY TERM

Longshore drift The transport of sediment along the coast.

Longshore drift is an irregular process and not the uniform one suggested in some diagrams. No two waves are identical and as a result sediment is moved in slightly different paths. The angle of wave approach may have a dominant direction but there will be times when the wind and therefore the waves come on to the beach from a different direction. Therefore, longshore drift is not a simple process. In the short term, on a day-to-day basis it is variable. In the longer term, over months and years, it operates in a preferred direction determined by the direction of prevailing winds. This preferred direction of movement is very important for long-term coastal erosion and deposition and therefore a significant factor in coastal management (page 185).

The operation of coastal sediment cells

The movement of sediment within the coastal zone is a very significant part of the coastal system. As human activity along coasts has grown, increasing attention is being paid to learning about sediment movements.

Coastal sediment budgets

Although difficulties remain as regards quantifying the various movements of materials, knowledge and understanding of how, where and when sediment moves within the coastal zone has greatly increased over the past three decades. **Sediment budgets** are used to assess what is happening to sediments within a specified location.

KEY TERMS

Sediment budget The balance of sediment volume entering and exiting a particular section of the coast.

Sink Anything that absorbs more of a substance than it releases.

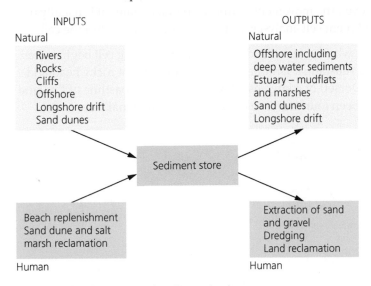

▲ **Figure 2.14** The coastal sediment budget

1 Balanced sediment budget

> volume of sediment in = volume of sediment stored + volume of sediment out

2 Unbalanced sediment budget

> volume of sediment in < volume of sediment stored + volume of sediment out

Human activities can have substantial impacts on sediment budgets, both positive and negative. While some actions are designed to have intended consequences for sediment budgets such as beach nourishment, other actions have unintended consequences. For example, in the past groyne construction reduced sediment flow, causing a net loss of sediment further along the coast. The construction of an inland dam can reduce a river's load and so less sediment enters the coastal zone.

Key components of a sediment budget are the sediment transfers. The direction and quantity of material moved indicates what changes to landforms might occur. Sediment is moved from one store or **sink** to another. Within the coastal system it is best to think of sinks as locations where deposition is prevalent, such as a beach, salt marsh or offshore bar. They can operate as both sinks and sources of sediment.

Sediment cells

KEY TERM

Sediment cell A stretch of coastline (including the nearshore area) within which sediment movement is largely self-contained.

Towards the end of the twentieth century, it was becoming clear that effective management of the coast required clearer knowledge and understanding of sediment movements. Research indicated that within the coastal zone sediment cells existed. The implication was that individual sediment cells (also known as littoral cells) essentially operate as closed systems. However, the movement of fine suspended material is usually not included in sediment cell analyses as it rarely settles out within the cell.

The boundaries where one cell ends and its neighbouring cell begins often coincide with a substantial landform such as a prominent rocky headland (Portland Bill, Dorset) or a large estuary (Severn). The coastline of England and Wales has been found to have eleven major or regional cells.

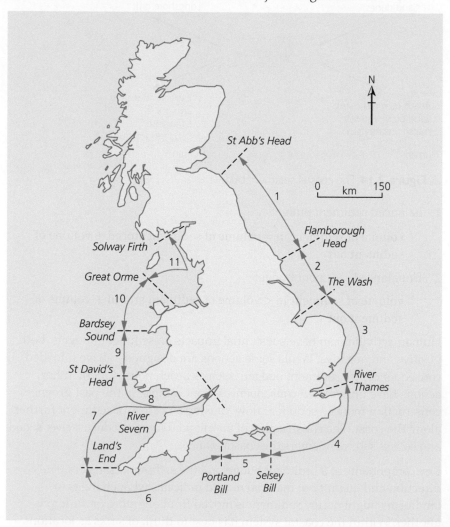

▶ **Figure 2.15** Regional sediment cells for England and Wales

Within this broad picture of regional cells there exist many sub-cells. For example, along part of the Jurassic Coast in Devon and Dorset, sub-cells operate between the Exe estuary and Portland Bill to the east and between the estuary and Start Point to the west.

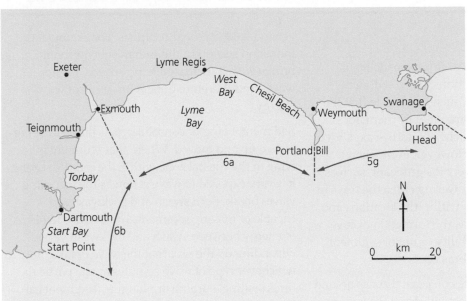

▲ **Figure 2.16** Sediment sub-cells along part of the Jurassic Coast

Within these sub-cells, even smaller-scale cells can be identified, such as a small cove. At the smaller-scales, however, sub-cells are more open as sediment moves in and out of them relatively freely.

The significance of time to the concept of sediment cells

The idea of sediment circulating within a defined stretch of coastline is an attractive one in the context of understanding the coastal system. However, sediment supply and transport is dynamic over a range of time scales. Consequently, a contemporary cell may not be fully understood in terms of 'average' conditions:

- Some rocky headlands are bypassed by sediment flows from one cell to another when especially vigorous storms occur.
- Material can be dragged offshore by the combination of exceptional wind and tide conditions.
- Locations that experience highly seasonal variations in factors such as energy inputs (wind and waves) and/or sediment inputs can see movements beyond cell boundaries.
- Some stores of sediment relate to processes operating in the geological past including during the Quaternary (see Chapter 5).

⑤ Evaluating the issue

▶ *To what extent is wave energy the most significant influence on sediment transport in the coastal zone?*

Context

Many coastlines have landforms made up of sediment. In some locations, these sediment-based landforms are of such large scale that the coastal landscape is dominated by them, for example the barrier beaches of north-east USA or Chesil Beach, Dorset. However, even where sediment seems relatively insignificant, such as on a rocky coast with towering cliffs, the ways in which sediment moves within the location can influence factors such as the formation of cove beaches and the availability or not of sediment for processes such as abrasion.

With mounting pressures on coastal zones around the world from human activities and sea level rise, the need to research how sediment is moved is vital. Coastal management decisions rely on authoritative understanding of coastal processes.

Evaluating the view that wave energy is the most significant influence

The crucial role of wave energy as an input into the coastal system is well-established. The transfer of energy from the atmosphere to the sea that waves represent is vital in the operation of so many processes within the coastal zone, not least of which is sediment transport. At the outset, it is important to appreciate that wave energy varies both spatially and temporally. Coasts in different parts of the world receive greatly contrasting amounts of wave energy, from high energy locations, such as much of north-west Europe, to low energy environments, such as West Africa. Variations in wave energy through time can be significant at locations experiencing

pronounced seasonal contrasts. For example, the monsoon season in the Indian subcontinent brings with it increased wind speeds, higher waves and therefore greater amounts of wave energy.

Sediment, being unconsolidated, is prone to being picked up and moved. Where that sediment is lying in the coastal zone within the intertidal zone, it is very exposed to wave action as the incoming waves break. Both swash and backwash are capable of moving sediment. The balance in terms of energy between swash and backwash varies depending on the way that the wave breaks. But whichever type of wave operates at any particular time, spilling, surging or plunging, sediment can be moved by the wave's energy.

With the movement of sediment comes erosion of that sediment. Processes such as abrasion, traction and saltation can lead to attrition of particles, making them smaller and more rounded. One consequence of this reduction in sediment calibre due to wave energy is that more sediment can therefore be transported by waves. Smaller-calibre sediment requires less energy to entrain and transport it, for example pebbles broken down into sand grains. This is an example of positive feedback, as decreasing sediment size due to wave energy allows more sediment to be transported which is then reduced in size even more and so becomes more susceptible to movement by moving water.

Even if the waves approach the shore parallel to the coastline, the relatively straightforward movement of sediment up and down a beach is significant. The landform known as a swash-aligned beach can be created but also considerable attrition of the sediment's individual particles can be carried out.

Where waves approach the shore at an angle, longshore drift of sediment occurs. This process, dependent on wave energy, is a significant component in the formation of landforms such as spits, zeta-form beaches, cuspate forelands, bars and barrier islands. Longshore drift is one of the processes most affected by coastal management. It has been poorly understood with the consequence that actions, such as groyne construction, have sometimes in the past achieved more harm than good. More focus is given to acquiring knowledge and understanding of wave-driven longshore drift in present-day coastal management projects.

One particular type of wave, the tsunami, possesses so much energy that it can severely affect a broader coastal strip. The scouring and transport of sediment can be on such a large scale that coastal landforms, such as beaches, are completely removed and sediment carried considerable distances inland and along the coast.

Evaluating the view that wave energy is *not* always the most significant influence

While wave energy clearly plays a major role in determining how sediment transfer processes operate, there are of course other factors to consider too. Sediment size (calibre) is a key influence on when, where and for how long sediment is transported. The fundamental relationship is that the larger the calibre of sediment, the more energy is required to move it. Pebbles and cobbles require more energy to be moved than sand grains. However, the relationship is not a simple linear one with the smallest sediment requiring the least energy, the largest sediment the most. One very significant point regarding sediment movement is that the smallest sediments, clays and silts, require *more* energy to start them moving than larger sand grains.

Along some stretches of coastline, river currents can play a significant role in transporting sediment. One aspect of this is the importance of river-derived sediment entering the coastal zone. Rivers in tropical and subtropical regions, such as the Amazon and Mekong, transport vast quantities of suspended sediment, which is the most significant source of sediment in the adjacent coastal systems. At smaller scales, such as the Mersey and Exe, and even locally where a stream enters the sea, river currents can be a significant source of energy transporting sediment.

Where a large tidal range exists, the exposure of a wide foreshore can lead to sediment drying out. Where this sediment is mainly sand, aeolian (wind) transport becomes very significant. Processes such as saltation and suspension can move substantial volumes of sand and are key in the formation of sand dune systems.

Earlier in the Quaternary, sea level was much lower than it is today as water from the hydrological cycle was 'locked up' as snow and ice on the land. As the ice age gradually came to a close, vast quantities of water began to flow from the land back to the seas. As sea level rose, sediment lying in areas exposed by the earlier fall in sea level was picked up and transported towards land. This 'historical' source of sediment is very significant in mid-latitude locations such as the British Isles. It is the case that wave energy was involved during this process. However, much of this sediment is 'relict' and not particularly actively transported by waves at the present time.

Arriving at an evidenced conclusion

The evidence would suggest that wave energy can probably be identified as the most significant influence on sediment transport in the coastal zone. Wave energy is, after all, so important in bringing energy into the coastal system that it underpins much landform development. It is important to focus on the role of breaking waves, as sediment on the seabed is not disturbed in water deeper than wave base. The inclusion of tsunami waves recognises the significance of extreme events in carrying work such as sediment transport.

Explaining how a landform or landscape has developed involves understanding how energy flows into, through and out of what is being studied, whether it be a large-scale barrier beach or an individual beach ridge. One approach has been to view landform/landscape development as the result of gradualism. In the context of the coastal zone, the ongoing receipt of energy, wave after wave, day after day, month after month and year after year, brought about slow, almost imperceptible development of features such as a beach.

With advances in databases and investigative techniques, geomorphologists have been able to identify extreme events which carried out extraordinary amounts of change in the environment in a very short period of time. It is now recognised that catastrophism plays a key role in geomorphology. Events, such as the very high energy earthquake off the coast of north-east Japan in 2011 and the consequent tsunami, may happen relatively infrequently (occurring once in several hundred years) but are responsible for a great deal of geomorphological change. Average

events, such as seasonal increases in wave energy, and frequent events, such as an individual wave, do bring energy into the coastal zone, but at levels that are comparatively small and therefore responsible for limited geomorphological change.

On the other hand, waves are not the only variable to be considered. Sediment size is an important element in the relationship between transport and energy, especially the need to consider the smallest calibre clays and silts. Additionally, wind energy can be very significant in some locations. The role of sea level change over geological time is another factor to be considered.

Although the relationship between wave energy and sediment transport seems at first sight to be straightforward, there are various factors to be evaluated. In the context of coastal management, priority must be given to understanding processes such as sediment transport before management solutions are proposed.

 KEY TERMS

Gradualism The theory that change, such as the development of coastal landforms and landscapes, is the cumulative product of slow but continuous processes.

Catastrophism The theory that change, such as the development of coastal landforms and landscapes, has resulted chiefly from sudden violent events.

Chapter summary

✔ There are three key marine erosional processes – hydraulic action, abrasion and attrition. In addition bioerosion can be a significant erosional element at some locations.

✔ Sub-aerial processes often make a substantial contribution to the development of landforms and landscapes in the coastal zone. Weathering processes can be classified as either physical or chemical and there are a range of types of mass movements operating at the coast. Different types of rock are more or less prone to different types of weathering and mass movement.

✔ Sediment, a major input into the coastal zone, has a variety of sources. The range in sediment size is important as it influences processes such as transport (of which there are four types) as well as the development of landforms.

✔ Both water and aeolian transport of sediment are influential in coastal landform development and can lead to substantial flows of sediment. On- and offshore transport of sediment can have significant influences on coastal landforms. Additionally, longshore movement of sediment is a key process in the development of landforms and landscapes of various scales.

✔ Coastal sediment budgets and the operation of sediment cells have become significant aspects in understanding processes and landforms in the coastal zone. Sediment circulation is a complex process involving a variety of factors, not all of which are active today. Past events and processes can be major factors in sediment supply and removal.

Refresher questions

1 Describe and explain the difference between the lithology and structure of rocks.

2 Describe the circumstances under which abrasion is likely to be particularly effective in the coastal zone.

3 Describe three ways in which bioerosion can operate in the coastal zone.

4 What is meant by the term 'sub-aerial weathering'?

5 Describe and explain the weathering processes most likely to affect granite and limestone.

6 Explain why water is such an important factor in mass movement.

7 Describe and explain the ways sand grains might be transported in the coastal zone.

8 What is meant by the term 'sediment cell'?

9 Outline the significance of past events in sediment supply in the coastal zone.

Discussion activities

1 With reference to the rock types in Figure 2.7, discuss the marine erosion processes likely to be responsible for the varying rates of cliff retreat.

2 Discuss why sediment origins and transport are a complex feature of the coastal system.

3 Consider the implications of currently occurring global warming in terms of possible changes to patterns of rainfall. Use your knowledge and understanding of the water cycle to inform your discussion. Discuss what could happen to sediment flows into and out of a stretch of coast if the area inland was exposed to either (a) an increase in rainfall or (b) a decrease in rainfall. Think about what impacts these changes in rainfall might have on rates of weathering and erosion of the landscape and on river flows.

4 Discuss the value of using sediment cells as a basis for understanding flows of material around a coastline. This would be valuable in the context of the real world. Research the arrangement of cells in a stretch of coastline you are studying. The website http://eprints.hrwallingford.co.uk has a dedicated Coast section at http://eprints. hrwallingford.co.uk/view/subjects/C.html which will help you navigate to maps of cells. Consult OS maps at 1:50 000 and 1:25 000 scales to help add detail to your discussions.

5 Look back over the section on mass movements. Research locations where slope instability is affecting coastal communities in terms of altering the physical characteristics of a place profile. Consider how the loss of homes, amenities such as coastal footpaths, coastal roads and/or railways affects both objective and subjective place meanings (a topic you are required to study as part of A-level Human Geography). The website of the British Geological Survey (www. bgs.ac.uk) offers a helpful starting point for information regarding landslips at coastal locations. You might then look at regional and local media websites covering such locations for more details of the impacts and reactions to coastal mass movements.

Further reading

Bray, M.J., Carter, D.J., Hooke, J.M. (1995) 'Littoral cell definition and sediment budgets for central southern England', *Journal of Coastal Research*, 11(2), pp.381–400

Davidson, T.M., Altieri, A.H., Ruiz, G.M., Torchin, M.E. (2018) 'Bioerosion in a changing world: a conceptual framework', *Ecology Letters*, 21(3), pp.422–38

Goudie, A.S., Brunsden, D. (1997) *Classic Landform Guide: East Dorset Coast*. Sheffield: Geographical Association

Goudie, A.S., Brunsden, D. (1997) *Classic Landform Guide: West Dorset Coast*. Sheffield: Geographical Association

Motyka, J.M., Brampton, A.H. (1993) *Coastal Management: Mapping of Littoral Cells*. HR Wallingford Report SR 328, Hydraulics Research Ltd, Wallingford, UK

Rosati, J.D. (2005) 'Concepts in sediment budgets', *Journal of Coastal Research*, 21(2), pp.307–22

CHAPTER 3

Contrasting coastlines

Coastlines are made up of a diverse range of landforms and landscapes. They can be viewed in plan or profile across the spatial scale from micro through to macro. They also exist across the temporal scale from minutes to millennia. And in many coastal locations, the influence of past processes and landforms continues to influence today's environment. This chapter:

- explores coastlines in plan
- investigates high energy coasts – marine cliffs and shore platforms
- investigates low energy coasts – beaches
- investigates low energy coasts – estuaries and deltas
- evaluates the extent to which marine cliffs are the result of erosion.

KEY CONCEPTS

Equilibrium The state of balance in a system. For example, on a marine cliff, when the volume of water entering the slope system is equal to that leaving, the slope tends to remain stable.

Feedback An automatic internal response to change in a system. For example, with undercutting of a marine cliff by waves, the slope angle of that cliff is steepened. The slope system is put under greater stress, making it more likely that mass movements will occur, a case of positive feedback bringing about disruption to the slope system.

Systems A system is a group of related objects. Physical systems tend to be open, having flows of energy and materials across their boundaries. A change in one part of the system brings about changes in other parts through the operation of feedback mechanisms. A beach operates as a system, with inputs of wave energy and sediment for example. The sediment moves according to the level of wave energy and can leave the beach system via longshore transport, for example.

Threshold A critical 'tipping point' in a system. For example, a beach can be considered as a system, with inputs, stores and processes, and outputs. If less sediment comes into a beach than is leaving it, then there will be a tipping point when the beach plan and/or profile will change.

1 Coastlines in plan

▶ *How and why do coastlines vary in their plan?*

Viewed from above, as on a map, the shape of the coastline becomes apparent. This **plan** view can be taken at any number of scales. At the global scale, the outlines of the continents reveal major features such as the

 KEY TERM

Plan View showing something in two dimensions. In the context of the coast it is usually taken as the view from above, a 'bird's-eye view'.

Gulf of Mexico. At this small scale, the details of a coastline's plan cannot be seen. Zooming in to the large scale, such as on an OS 1:25 000 map, individual beaches become clear. An even closer look reveals features such as small indentations to a cliff.

Concordant and discordant coasts

Two contrasting coastline plans can be identified:

● Concordant   ● Discordant.

Variations in plan are strongly influenced by geology. Lithology and structure affect the ways weathering, erosion and mass movement operate. Together they mean that **differential denudation** operates.

A **concordant** plan is one where the coastline is relatively straight with no large bays and headlands. The principal influence on this shape is geology, with the rocks arranged parallel to the coastline. The western coastlines of both North and South America are considered concordant for much of their length. Altering the scale to look at a smaller coastal area, the south-west and northern stretches of the Gower Peninsula, South Wales are essentially concordant (Figure 3.1).

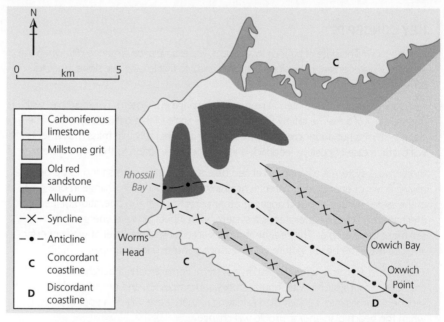

► Figure 3.1 The relationship between coastal plan and geology: Gower Peninsula, South Wales

A **discordant** plan is one possessing significant bays and headlands. The north east coast of the USA is an example, with Cape Cod and Long Island located perpendicular to the overall trend of the coast. On the Gower Peninsula's south east section, substantial bays and headlands exist.

Along concordant coastlines, the geology tends to lie in a parallel trend to the coastline. If the rock forming the cliff line is particularly resistant, coves (small bays) and bays can form once a breach has been made. Either the sea or a river can make this opening. Marine and sub-aerial processes can then attack the less resistant rock just inland (page 127).

Discordant coastlines develop where geology trends perpendicularly to the coastline. They are also influenced by the way that **synclines** can be worn down to form valleys which can become coastal bays. **Anticlines** tend to form ridges which become headlands at the coast.

The plan view of a coastline is dynamic through time. Over geological time, coastlines have altered substantially. Just going back some 120 000 years, sea level was much lower than today, perhaps about 120 metres, resulting in a different look to the British Isles. No English Channel existed and the area of the North Sea was largely dry land and swampy areas. Today, the ebb and flow of the tide alters the plan of a coastline. One of the Channel Islands, Jersey, experiences a 40 per cent increase in its area at low tide and has a very different plan view.

> 🔑 **KEY TERMS**
>
> **Syncline** A downfold in rock strata.
>
> **Anticline** Where rock strata are folded upwards.

CONTEMPORARY CASE STUDY: THE JURASSIC COAST, DORSET

Along this coastline, the more resistant Jurassic limestones (Portland and Purbeck limestones) have been folded so that they dip very steeply landwards. In places they continue to resist wave erosion and weathering but elsewhere they have been broken through. This has allowed the less resistant Greensand and Wealden sands and clays to be worn away forming bays, such as Lulworth Cove and St Oswald's Bay. Moving inland, the next rock is a limestone, chalk which is able to form near vertical cliffs.

It is important to appreciate that marine processes are not solely responsible for this pattern. Tectonic forces crumpled the rocks of what is now Dorset some 65 Ma (million years) BP when the African and Eurasian Plates collided.

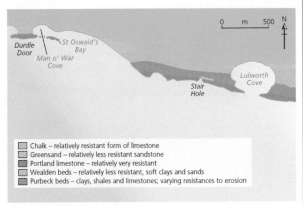

Chalk – relatively resistant form of limestone
Greensand – relatively less resistant sandstone
Portland limestone – relatively very resistant
Wealden beds – relatively less resistant, soft clays and sands
Purbeck beds – clays, shales and limestones; varying resistances to erosion

▲ **Figure 3.3** The geology of part of the Jurassic Coast, Dorset

The effects of folding can be very localised, with variations in angle of dip of between 45° and 90° along the Dorset coast. In addition, the most recent ice age has influenced the landforms and landscapes. During the ice age this part of the British Isles was not covered by ice but was intensely cold with the land frozen to great depth and sea level some 120 metres lower than today. As the ice receded and the frozen land thawed, vast quantities of water were released from ice melt. Fast-flowing rivers made their way to the sea, cutting through all the rocks, including the resistant Jurassic limestones. This helped the sea break through the resistant rocks and form coves and bays. As sea level gradually rose, eventually wave action began to operate on the rocks, producing today's recognisable landscape.

This sequence highlights the importance of recognising the interaction of past and present-day processes and events when interpreting a coastline.

▲ **Figure 3.2** Concordant coastline, Jurassic Coast, Dorset, looking eastwards across Man o' War Cove and St Oswald's Bay

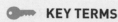

Another type of concordant coastline is Haff coastlines, which are found in the southern Baltic Sea. The name refers to the shallow **lagoons** separated from the sea by long, narrow sand spits topped with dune systems that characterise this area. Unlike other locations, it is deposition of sediments that has created the concordant plan, such as in Lithuania and Poland.

② High energy coasts – marine cliffs and shore platforms

▶ *What factors interact to form marine cliffs and shore platforms?*

Steep slopes – cliffs – along the coastline vary in both their plan and their **profile**. Thinking about cliffs as a system helps us understand why the **morphology** of cliffs varies from one place to another (Figure 3.4).

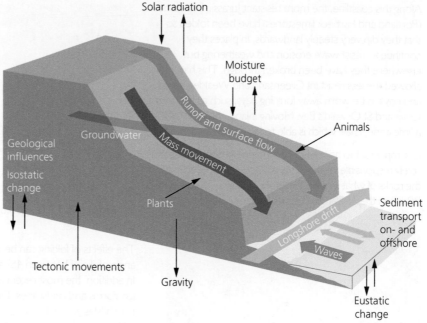

▲ **Figure 3.4** The cliff system

Any change in one or more of the components of the cliff system results in changes to the cliff's morphology.

Cliff profiles and plans – the role of geology

The primary influence on the development of coastal cliffs tends to be their lithology and structure. The contrast in resistance to weathering and erosion broadly coincides with the three main categories of rocks:

- Igneous – formed from previously molten material that either cooled beneath the surface and is now exposed at the surface (granite), or erupted and then cooled at the surface (basalt).

- Sedimentary – formed from the fragments and particles of older rocks which have been weathered and eroded. The sediment is then deposited by wind or water, often as identifiable layers or strata (clays and sandstones). Some sedimentary rocks have organic elements, for example shells (chalk).
- Metamorphic – rocks that have been significantly altered through the actions of heat and pressure (mudstone → slate; granite → gneiss; limestone → marble).

The influence of lithology

Most igneous and metamorphic rocks have lithologies resistant enough to form steep cliffs. Some sedimentary rocks, for example Carboniferous Limestone, Old Red Sandstone and chalk, also form steep cliffs.

In contrast, unconsolidated rocks such as clays and recently deposited sands tend to result in low-angled slopes. However, local conditions can result in steep cliffs in relatively weak rocks. Constant marine erosion at the cliff base leads both to slope failure and mass movement of material on to the beach. Clays, sands and **glacial till** are soon broken down and transported by waves and currents, leaving a steep slope to be undercut again.

The influence of structure

Geological structure involves a number of features:

- joints and bedding planes
- angle of dip of strata
- faults.

Whether joints and/or bedding planes are present and at what density greatly affects the susceptibility of a cliff face to weathering and erosion. Increased water access accelerates processes such as salt crystal growth, hydrolysis and freeze-thaw (page 32), as well as allowing wave attack to penetrate further into the cliff face. Rock with a well-developed joint and bedding plane pattern tends to form steep cliffs.

At the cliff base, wave attack creates an overhang. This will eventually collapse, exposing fresh rock at the cliff face which has retreated a little. The debris at the cliff foot is further broken down until small enough to be transported away by

▲ **Figure 3.5** Steep granite cliff, Jersey; because of its castle-like appearance, this type of cliff is known as castellated

▲ **Figure 3.6** Steep cliffs in Triassic sandstone and mudstone, Sidmouth, Devon

 KEY TERM

Glacial till A general term covering all the materials deposited directly by ice, usually a mixture of clays, sands and rock fragments.

▲ **Figure 3.7** Vertical cliffs in horizontally bedded chalk, Étretat, north-west France

the sea. The cliff base is once again subject to wave attack and the cycle continues.

The angle of dip of rocks, especially sedimentary ones, is a significant influence (Figures 3.6, 3.7 and 3.8).

- Cliffs with horizontal strata retreat parallel to their faces as undercutting leads to rockfall and toppling.
- Where the strata dip seaward, the cliff face tends to maintain an angle similar to that of the dip as loosened blocks slide down to the cliff foot (Figure 2.5).
- Where the strata dip landwards the profile tends to be relatively stable. Often such slopes are slightly convex because marine processes are less effective than the sub-aerial attack on the upper portion of the cliff. The average angle of the joint pattern also influences cliff shape.

(a) Uniform horizontal strata produce steep cliffs

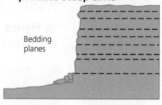

Bedding planes

(b) Rocks dip gently seawards with near-vertical joints

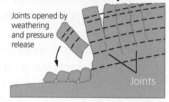

Joints opened by weathering and pressure release

Joints

(c) Steep seaward dip

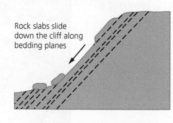

Rock slabs slide down the cliff along bedding planes

(d) Rocks dip inland, producing a stable, steep cliff profile

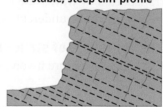

(e) Rocks dip inland but with well-developed joints at right angles to bedding planes

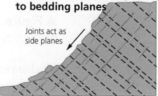

Joints act as side planes

(f) Slope-over-wall cliffs

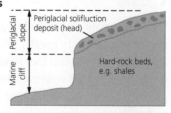

Periglacial slope

Periglacial solifluction deposit (head)

Marine cliff

Hard-rock beds, e.g. shales

▶ **Figure 3.8** Cliff profiles: the influence of angle of dip

At locations where cliffs contain more than one rock type, profiles are more complex. A characteristic profile found in areas where either ice sheets left till sheets behind or where periglacial processes were active is known as a slope-over-wall profile, with a steep lower portion and a more gentle upper part. Active marine erosion and transport creates the lower profile while sub-aerial processes, both present-day and those operating in the past, result in upper-profile slopes of about 25° to 30° (Figure 3.8f).

Small-scale coastal plan and profile

When investigating a stretch of coastline in detail, small-scale features become clear. In some locations these can be due to areas of relative weakness within a single rock type. For example, joints, bedding planes and faults are where rock is prone to weathering and erosion processes.

Small-scale coves and headlands are likely to owe their origin, in part, to differential denudation. A **geo** is a deep, narrow cleft that follows a line of weakness inland (Figure 3.9). It may have begun as a long and narrow cave but then its roof collapsed.

Within a cliff face, variations in rock type can lead to small-scale ridges and indentations. This can also result from differences in the lithology of the same rock type (Figure 3.10).

🔑 **KEY TERMS**

Periglacial Literally means on the fringe or close to an ice sheet or glacier. It also includes high latitude locations.

Geo A deep, narrow inlet on a cliffed coastline.

▲ **Figure 3.9** Geo cut along a weakness in resistant Silurian sediments, Valentia Island, Ireland

▲ **Figure 3.10** Alternating strata of contrasting degrees of resistance in Jurassic sandstones, West Bay, Dorset

Cliff morphology and the balance between marine and sub-aerial processes

Marine cliffs are essentially slopes at the seaside. These slopes, unlike ones inland, experience marine processes towards the bottom of the slope. The relative balance, therefore, between marine and sub-aerial processes is significant as an influence on cliff shape.

Marine erosion attacks the cliff between high and low tide levels. Debris from the eroded cliff, as well as material produced by sub-aerial processes operating higher up the cliff face, falls to the cliff base. Marine processes then remove this sediment.

It is perhaps helpful to think of three scenarios in terms of the relative balance between marine and sub-aerial processes (Figure 3.11):

(a) Marine processes > sub-aerial processes
(b) Marine processes = sub-aerial processes
(c) Marine processes < sub-aerial processes.

(a) Marine processes more effective than sub-aerial processes

Relatively resistant rocks. Cliff shape mainly the result of geological structure.

Comparatively slow sub-aerial denudation

Mass movements mainly rockfall

Comparatively rapid marine erosion

Wave-cut notch

Marine removal of debris offshore and longshore

e.g. chalk cliffs

Mean sea level

(b) Marine processes and sub-aerial processes in balance

Sub-aerial denudation at moderate rates

Mass movements mainly rockfall, sliding and slumping

Moderately resistant rocks

Episodes of marine erosion and cliff collapse form a beach. Inputs from the cliff balance with sediment removed by marine action.

Marine removal of debris offshore and longshore

e.g. wide variety of rock type

Beach

Mean sea level

(c) Marine processes less effective than sub-aerial processes

Sub-aerial denudation at rapid rates

Mass movements mainly sliding, slumping and flows

Inputs of cliff material occur at a faster rate than marine erosion

Occasional high wave energy removes debris

Weak rocks due to lithology and structure

Marine removal of debris offshore and longshore

e.g. clays and sands

Mean sea level

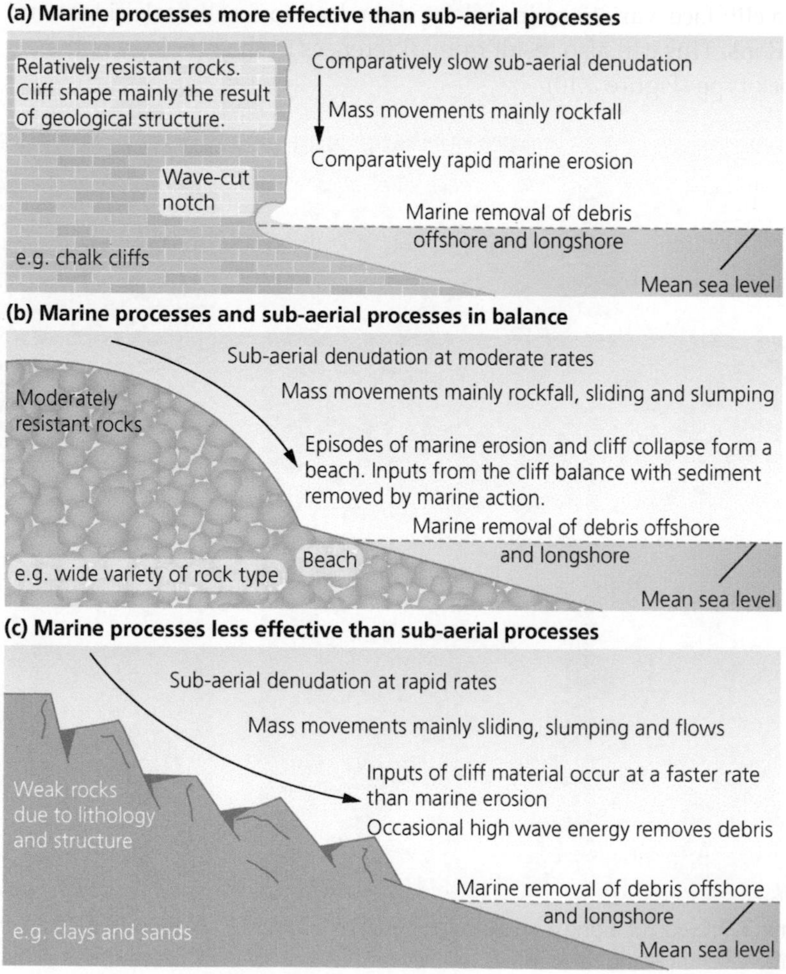

▶ **Figure 3.11** The relationship between sub-aerial and marine processes and the effect on cliff shape

Under (a), the cliff is under constant attack from wave action and any debris is soon removed. This tends to result in a steep cliff profile. Under (b), wave action is not as effective as sub-aerial processes and so debris weathered from the upper cliff face accumulates at the bottom of the cliff. The upper cliff face is worn back while the cliff is protected at its base. The cliff profile may have a gentler slope in its upper section. In (c) the cliff shape reflects factors such as rock lithology and structure. It can often be a fairly chaotic profile that can change rapidly depending on factors such as weather and wave energy. A prolonged period of rain, for example, can lead to so much water entering the cliff system that the slope fails as a rotational slide or a slump.

The influence of latitude on the balance between marine and sub-aerial processes

It has been suggested that cliff morphology varies with latitude because the relative rates of debris supply and removal are related to the coastal energy system. Wave energy is highest along mid-latitude coasts (30° to 60°) and lowest in the low latitudes (tropics). In the high latitudes (e.g. Arctic Circle), coasts tend either to be relatively sheltered or to experience ice in winter and, therefore, receive low wave energy. Debris removal and supply from marine processes are at a maximum in the mid-latitudes.

Variations in sub-aerial influences are more complex. Humid, low-latitude areas, such as West Africa and South-East Asia, tend to have slopes covered by vegetation, which yield relatively small amounts of sediment falling off cliffs. However, arid tropical coasts, such as Namibia, south-west Africa or around the Arabian Peninsula, lack vegetation cover and so steep cliffs are more common. Weathering processes such as frost shattering are active in high latitude locations. The large volumes of scree produced are incapable of being removed by the low wave energy. In locations such as Spitzbergen or northern Alaska, the slopes reflect the angle of rest of the sediment that makes up the scree.

Features of cliff retreat: plan and profile

Where there is a clear divide between the sea and the land, for example along coastlines in which the land surface is significantly above sea level, a variety of distinctive erosional landforms can develop.

A notch forms around the mean high water mark. It extends under the cliff face as a result of a combination of wave erosion, bioerosion and weathering. Notches are found in different geologies and are particularly common in limestones.

Sea caves develop around the mean water level and extend into the cliff base. In scale they can reach substantial size, some tens of metres deep and several metres tall. They often follow weaknesses in the rock such as an area of dense jointing or a fault. Where a vertical weakness exists, a blowhole can form, extending through the roof of the cave and opening on to the cliff top.

 KEY TERMS

Scree Broken down rock fragments, usually quite angular. Also known as talus.

Notch An indentation at the base of a cliff.

Sea caves Voids of various sizes in solid rock. They are common along cliffed coasts where the rock has structural weaknesses such as joints, bedding planes and faults.

Blowhole A vertical shaft leading from a sea cave to the surface.

▲ **Figure 3.12** Spouting Horn, Kaua'i Island, Hawaii. This blowhole follows the line of an eroded lava tube. The column of water is about 15 metres tall

▲ **Figure 3.13** Arch cut in resistant Silurian sediments, Valentia Island, south-west Ireland

▲ **Figure 3.14** Stacks and stumps in resistant Devonian slates and sandstones, Bedruthan, North Cornwall

When the tide is high enough, water can be forced through this natural tube to give a spectacular release of water at the cliff top.

At locations where small-scale headlands protrude from the general line of cliffs, wave erosion can be focused on these locations by wave refraction. Notches and caves can form which can extend through the headland to give an **arch**. These features can last for many decades, but eventually they collapse to leave an isolated column of rock known as a **stack**. Stacks continue to be worn away until a **stump** is left as a small platform of rock, sometimes covered at high tide.

Shore platforms

Shore platforms are horizontal or gently sloping landforms extending seawards from cliffs. They are relatively flat rock surfaces, with rock pools and occasional larger blocks of rock. High tides tend to cover them; they are exposed at low tide.

▶ **Figure 3.15** Shore platform in Carboniferous sandstone, Northumbria

60

It is clear that shore platforms are intrinsically linked with cliff retreat (Figure 3.16). However, it would be incorrect to assume that the same processes created both cliffs and platforms.

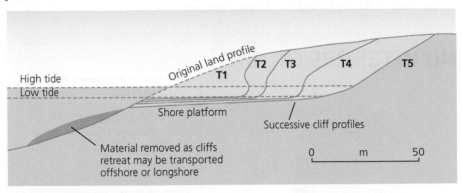

▲ **Figure 3.16** Cliff retreat and shore platform development

Wave-cut platform was the term originally used to indicate the assumed predominance of wave action in the production of this landform. However, it is now recognised that it is the interaction of several processes that are responsible and so this term is misleading. It is also appreciated that the relative significance of various processes varies in different parts of the world. Shore platforms in sheltered locations are unlikely to have been carved solely by wave action. Where wave energy is high, the role of weathering in partially weakening rock before the waves complete the wearing away, should not be underestimated. In particular, wetting and drying and salt weathering can be very effective on shore platforms. In some locations, bioerosion plays an important role. The debate continues regarding the relative importance of processes responsible for shore platform development.

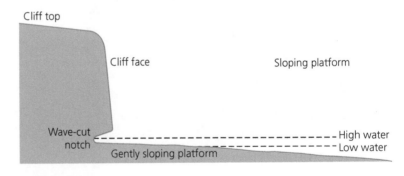

Two basic types of shore platform tend to be recognised (Figure 3.17):

● Sloping – gradient 1° to 5°, a continuous platform without major interruptions in the slope
● Sub-horizontal – gradient almost flat with a small rampart and low cliff at the low water mark.

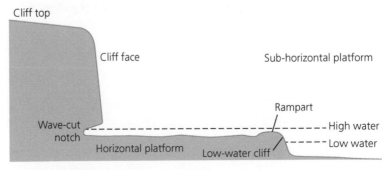

▲ **Figure 3.17** Types of shore platform

Some researchers suggest that sloping platforms are common in macro-tidal environments such as the British Isles. Sub-horizontal platforms are more likely to occur along coasts with meso- and micro-tidal regimes, such as much of Australia and the Mediterranean.

ANALYSIS AND INTERPRETATION

Study Figure 3.18.

▲ **Figure 3.18** Cliffs in landward dipping chalk, Dorset

(a) With reference to Figure 3.18, explain the influence of geology on the cliff profile.

GUIDANCE

The term geology includes both lithology and structure. The former refers to the physical and chemical composition of a rock, the latter to features such as bedding planes, joints and folds. The steep profile indicates that the cliff face is undercut at its base and then falls down. Because the chalk strata dip down towards the land, forces within the chalk act to keep its profile relatively stable. If they dipped in the opposite direction then rock would be more likely to slip down towards the sea. The joints and bedding planes within the chalk are lines of relative weakness. In the foreground there is a line of small caves and on the far headland a larger cave. These might indicate areas within the chalk where it is less resistant and so the profile is interrupted by these features.

(b) With reference to Figure 3.18, suggest why there is little debris from the cliff at its base.

GUIDANCE

Because of the steep profile, it is reasonable to expect that rock fall occurs quite often so that the almost vertical angle is maintained. If this is the case, what then has happened to the debris that has fallen off the cliff face? One obvious explanation lies with wave energy. The small beach suggests that wave energy regularly reaches the bottom of the cliff, allowing erosional processes such as hydraulic action, abrasion and attrition to work away at chalk debris. Once the chalk has been broken into small particles, it can be transported away by the sea. In addition, the chalk will break down chemically and can be washed away in solution.

In addition it is worth pointing out that the cliffs in the foreground have significant proportions of the cliff face covered by vegetation. This can only gain a roothold if the chalk is relatively stable. There may be sections of these cliffs that are not that active and so not experiencing a great deal of rock fall.

(c) Examine the role sub-aerial processes can play in the formation of marine cliffs.

GUIDANCE

All marine cliffs experience two sets of processes wearing them down: marine (waves) and sub-aerial. The latter involves any of the weathering processes, such as carbonation and hydrolysis as well as erosion by water running over the cliff surface. The term also includes mass movements. Where marine processes are very active, such as in high wave energy locations, cliff erosion can be an active process. Cliff retreat can be relatively rapid in geomorphological terms and result in the cliff profile being maintained at generally steep angles.

In locations where sub-aerial processes are either equal to or more effective than marine processes, cliffs tend to have a profile that is generally more rounded. The upper part of a cliff is worn way by weathering and erosion by flowing water, while the lower part of the profile resists the attack of the sea.

On geology that is particularly susceptible to sub-aerial attack and mass movement, slope failures such as slumps, slides and flows can produce an overall lower-angled profile. However, in the case of rotational slips, some cliff sections can remain quite steep until further denudation occurs. And at the bottom of the cliff, wave action usually quickly removes any debris, creating a steep lower portion.

 # Low energy coasts – beaches in plan and profile

▶ *What factors interact to form beaches?*

The landforms created by sediment deposition can be just as spectacular as those found along rocky coasts. And just as with cliffs, low energy landforms have subtlety in both their formative processes and shapes. The most widespread category of depositional landform is the beach.

Beaches are stores of loose sediment within the coastal system. A variety of unconsolidated material, most commonly sand and shingle, can accumulate between the area where waves begin to experience friction with the seabed and the zone landwards of the high tide level. Beaches are remarkably dynamic landforms, experiencing almost continual changes in shape. Sediment responds rapidly to changing inputs of energy from wind, waves, currents and tides. Many beaches exist in a state of dynamic equilibrium. There are also occasions when dramatic changes occur, particularly in response to abrupt increases in wave energy, for example during a high energy storm – a state of meta-stable equilibrium (pages 20–1).

In the mid latitudes, pebble beaches are most common; in the low latitudes, sand beaches prevail.

Beach plan: large-scale landforms

Beaches tend to follow the general trend of the coast. However, some extend out from the coastline. The main influence on beach plan, the bird's-eye view, is wave energy and in particular the relationship with the prevailing wave direction. There are two fundamentally different types of beach plan:

- Swash-aligned – beach orientated parallel to the shoreline
- Drift-aligned – beach orientated obliquely to the shoreline.

A swash-aligned beach is usually a fairly closed sediment system as there is limited transfer of material into or out of the beach. Wave refraction causes wave crests to curve to the shape of the shoreline with the paths of swash and backwash the same. Longshore transport of sediment is minimal and deposition predominates.

Some swash-aligned features are large scale, such as Rhossili Bay, Gower, South Wales which is some 5 km long. Grève de Lecq

▲ **Figure 3.19** Crescent-shaped, swash-aligned beach plan, Grève de Lecq, Jersey

(Figure 3.19) is about 400 metres in length, while a cove may have a small beach of just a few metres nestled up against the cliff base.

A drift-aligned beach is a more open system. Sediment enters at one end, passes along the length of the beach due to longshore drift and then leaves the beach at the other end. The predominant angle of wave approach brings with it enough energy to keep the sediment moving.

There is one type of beach plan that seems to be the result of both swash and drift processes. Where headlands partially interrupt longshore drift, **zeta-form** or **fish-hook** beaches develop.

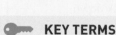

KEY TERMS

Swash-aligned Beaches that are crescent-shaped accumulations resulting from fully refracted waves.

Drift-aligned Beaches that tend to be longitudinal accumulations resulting from incomplete wave refraction.

Zeta-form or **fish-hook** Beaches that are relatively narrow at one end and wider at the other.

Refraction around the headland produces wave crests that have a small angle of approach to the shore. Further along the bay the angle of wave approach is greater, which allows longshore transport of sediment.

Spits

Where a beach accumulates out from the shore in the direction of longshore drift, a **spit** usually forms. Where the spit joins the land is known as the proximal end, while the end projecting out from the coast is the distal end. They are more commonly found in locations where:

- there is a ready supply of sediment, particularly sand and shingle
- longshore drift is active
- the coast has an abrupt change in direction, such as at an estuary or bay
- tidal range is limited, generally < 3 metres.

When the tidal range is relatively small and there is sufficient wave energy, sediment is readily transported. If the supply of sediment ceases or is substantially reduced, the spit is likely to lose its sediment and eventually disappear.

Where the longshore current enters deeper water near the distal end, wave energy is dispersed throughout more water. Rates of transport are reduced and deposition increases. Waves refract around the end of the spit, taking sediment with them. In the relatively low-energy conditions just behind the distal end of the spit, sediment builds up in the form of a **recurve** (Figure 3.21). Spits with several recurves along their length can be termed compound spits. It is likely that where a recurve is pronounced, waves approaching from a different direction to the main trend in drift bring sufficient energy to move sediment in that different direction. Recurves can also mark different phases in the development of a spit. They can be

Beach plan is drift-aligned as waves approach at oblique angle

Headland

Longshore drift

Beach plan is swash-aligned as waves refract round headland

Headland

Prevailing wave advance

▲ **Figure 3.20** Zeta-form beach – form and processes

🔑 KEY TERMS

Spit A long, narrow beach joined to the mainland at one end only.

Recurves (sometimes termed recurve laterals) Hook-like ridges that develop towards the distal end of a spit. They are mostly found made of shingle.

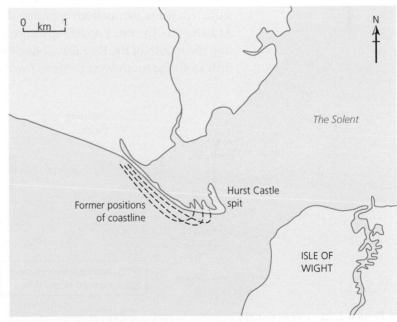

0   km   1

N

The Solent

Hurst Castle spit

Former positions of coastline

ISLE OF WIGHT

▲ **Figure 3.21** Recurves on Hurst Castle spit, Hampshire

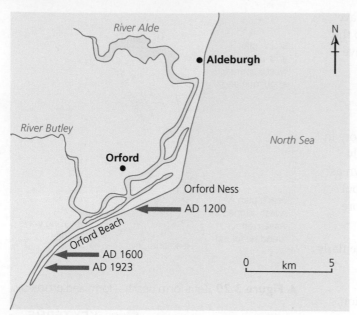

▲ **Figure 3.22** River mouth diversion due to spit growth

associated with periods when the coastline, including the spit, was seaward of its later position.

Most spits have formed since sea level stabilised about 4000 years ago following the rise in sea level resulting from the end of the ice age. Spits are landforms that adjust quickly to changing inputs of energy and sediment. In the course of a couple of centuries, a spit can extend across the mouth of a river to divert its exit to the sea by several kilometres. For example, in Suffolk, the exit of the River Alde into the North Sea has been diverted south by some 12 km due to the growth of Orford Ness spit (Figure 3.22).

A pair of spits facing each other either side of an indentation in the coastline might seem contradictory. At Christchurch Harbour, Dorset, a spit once stretched from Hengistbury Head in the south almost as far as Highcliffe Castle to the north-east (Figure 3.23). Violent easterly storms breached the spit in 1886 and in 1935. Evidence of drift from the north is unconvincing, so it seems that the double spit was once a single feature formed by longshore movement of sediment from the south. This is an example of where past events help understand present-day landforms. In some locations, however, there is sufficient wave energy from opposing directions to generate double spits. At Exmouth, Devon, a double spit once extended from the west and east into the mouth of the Rive Exe. Today, only the spit formed by longshore drift from the south-west exists as Dawlish Warren (pages 197–9). The spit on the opposite bank is now part of the built-up area of Exmouth following considerable land reclamation and the construction of a substantial sea wall. This stretch of coast still experiences sufficient occasional wind energy from the east and south-east to move some of the town's beach sediment east to west. This acts as a reminder of how the 'lost' spit was developed.

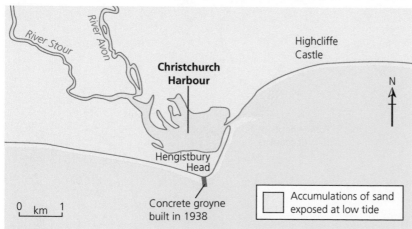

▲ **Figure 3.23** Double spit at Christchurch harbour, Dorset

ANALYSIS AND INTERPRETATION

Study Figure 3.24.

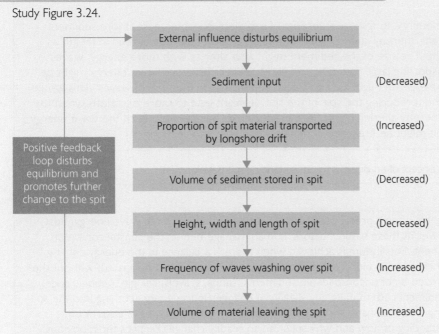

▲ **Figure 3.24** Positive feedback operating within a spit system

(a) Using Figure 3.24, explain how two external influences might cause the spit's sediment supply to be disturbed.

GUIDANCE

Disturbance to the supply of sediment could happen either naturally or more likely due to human factors. If a cliff was acting as a source of sediment due to its weathering and erosion by sub-aerial and marine forces, it could end up receding beyond the reach of wave action. The supply of sediment from the cliff would then be reduced. The sediment might be mainly from a river. If land-use changes in the river's drainage basin, this might reduce sediment in the coastal zone. **Afforestation** would protect the soil from erosion and transport by overland flow. Less sediment therefore enters the channels draining the basin and less sediment is carried down into the coastal zone. Human intervention at the coast could reduce sediment supply. Cliff protection might stabilise the cliff face and prevent sediment falling off and entering the sea. The erection of groynes along the coast could stop or reduce longshore sediment flows.

(b) Suggest what possible changes might occur to the spit if wave energy increased.

GUIDANCE

With an increase in wave energy, longshore transport of sediment might increase. More sediment would enter the spit system, thereby adding to its volume. The spit could well extend in length as well as increasing in height. The size of the sediment might also alter as with more energy, waves would be capable of transporting larger-sized sediment. However, increased wave energy might be destructive of the spit as backwash might exceed swash depending on the type of waves. This would lead to a net loss in sediment reducing the size of the spit. It might lead to more occasions when the waves overtop the spit, thereby altering its profile by reducing its height in places. If the wave energy was great enough, the spit might be breached completely and split into sections.

(c) Explain the value of the concept of equilibrium to understanding spit development.

GUIDANCE

The very existence of a spit is often a precarious balance between inputs of wind, wave energy, tidal energy and sediment. As well as these natural factors, human management in the coastal zone is an increasingly active component. Equilibrium is achieved when there is a balance in the energy coming into a particular stretch of coast or an individual landform such as a spit and that dissipated, without the coast changing. Spits are made up of unconsolidated sediments, sands, gravels, shingle, pebbles and cobbles. Because these sediments are mobile, any significant changes in the energy coming into the spit will soon bring about changes to the shape and size of the spit. The different types of equilibrium (page 20) can help understanding of how and why spits can so readily change. Perhaps most obviously, sudden dramatic high energy events such as a storm bring vast amounts of energy to the coast, possibly resulting in a severe reduction to a spit's height and/or its breaching. This would be an example of meta-stable equilibrium as the spit system would then need to adjust to the new situation.

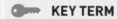

KEY TERM

Afforestation The planting of trees.

Cuspate forelands and tombolos

Cuspate forelands vary in scale but are not particularly small. Dungeness foreland extends some 30 km along the Kent coast while projecting about 15 km into the English Channel. Along the Carolina coast, USA, cuspate forelands can reach 150–200 km along the side attached to the mainland, such as Cape Fear (page 70).

Cuspate forelands seem to represent locations where sediment, transported by longshore drift, becomes trapped when equilibrium between sediment inputs and energy to move them is reached. However, one explanation does not account for all such landforms. Where there is an offshore island, wave refraction around it on either side brings waves, travelling in opposing directions, together in the shadow of the island. Sediment is deposited in this location. At some places this process leads to a **tombolo** forming. The town of Llandudno, North Wales, is built on a tombolo linking the mainland with the Great Orme, a former offshore island.

KEY TERMS

Cuspate foreland A triangular-shaped projection attached to the mainland with its apex pointing out to sea.

Tombolo A bar or spit connecting the mainland to an island.

CONTEMPORARY CASE STUDY: DUNGENESS

Dungeness, a broad expanse of shingle ridges in Kent, seems to have existed first as a spit that extended eastwards due to longshore drift. Rivers brought sediment from inland, which added to that already in the coastal zone, much of which probably came from the post-glacial rise in sea level. Wave energy from the east may also have had a role in promoting sediment accumulation at this location. In the thirteenth century, the spit was breached at its proximal end. From then on sediment eroded from the south-west part of the coast was deposited further east as a series of shingle ridges building out from the shore. The apex was able to build out from the shore as the short fetch to the south-east meant only limited wave energy came into the area. Human activity was another factor as drainage of marshland, changes in land use and coastal protection modified natural processes.

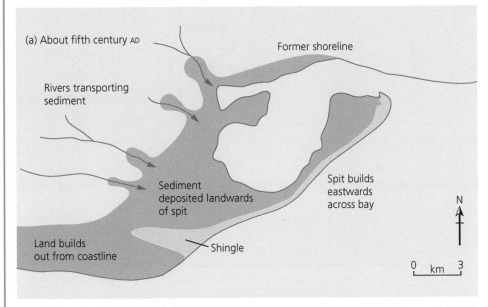

(a) About fifth century AD

Former shoreline

Rivers transporting sediment

Sediment deposited landwards of spit

Spit builds eastwards across bay

Land builds out from coastline

Shingle

0 km 3

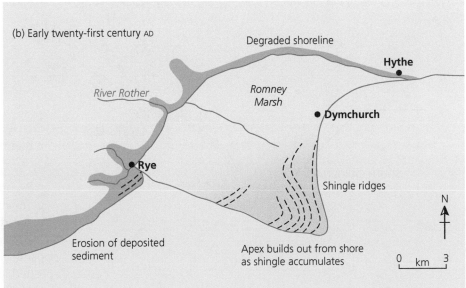

(b) Early twenty-first century AD

Degraded shoreline

River Rother

Romney Marsh

Hythe

Dymchurch

Rye

Shingle ridges

Erosion of deposited sediment

Apex builds out from shore as shingle accumulates

0 km 3

◄ **Figure 3.25**
The formation of Dungeness

Bars and barrier islands

A **bar** is a generic term given to a range of sediment accumulations in the coastal zone. They range in scale from comparatively small features just a few metres wide and a couple of hundred metres long to landforms over 1 km wide, hundreds of kilometres long and up to 100 metres in height. At this larger scale they are known as **barrier islands**. About 10–15 per cent of the world's coastlines are made up of barrier islands, being particularly common in low- to mid-latitudes. The east coast of the USA has some well-developed barrier island systems (Figure 3.26).

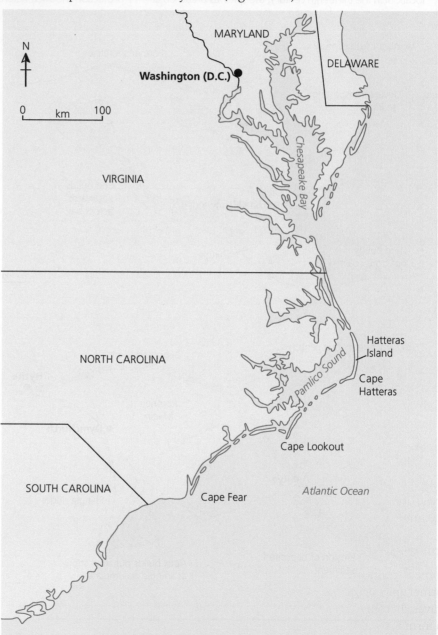
▶ **Figure 3.26** Barrier islands, eastern USA

Factors common to barrier islands are:

- a gently sloping offshore gradient
- limited tidal range < 3 m
- relatively high wave energy.

Most barrier islands have a series of dune ridges on their seaward face. Large quantities of sediment are transported daily along and around barrier beaches. Where the barrier is made of coarser sediments such as shingle, water can seep through the barrier into the lagoon behind it. Tidal currents and the flows of river and lagoon water combine to produce dynamic patterns of erosion and deposition along a barrier.

The formation of bars and barrier beaches Controversy continues to surround how bars and barrier beaches were formed; it is likely that alternative theories apply at different locations and times.

Changing sea level

Cores obtained from barrier islands along the east coast of the USA contain deposits of silt and clay beneath the barrier's sand and shingle. The presence of these small-calibre sediments suggests that they formed under low-energy conditions such as estuaries or lagoons trapped behind the barrier. It seems that barriers rolled landwards as sea level rose at the end of the last ice age (Figure 3.27).

Another theory suggests that offshore bars developed when sea level was a little higher than it is today. When sea level reduced, these accumulations were exposed and became barrier islands (Figure 3.28).

Spit breaching

Some barriers are thought to be the result of former spits being breached by high energy waves that then eroded a permanent channel to form islands (Figure 3.29).

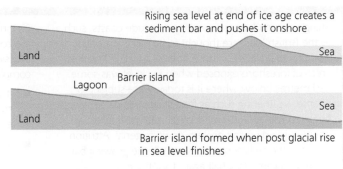

▲ **Figure 3.27** Barrier island formation – rising sea level

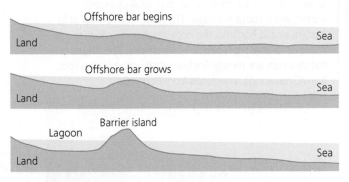

▲ **Figure 3.28** Barrier island formation – reduced sea level

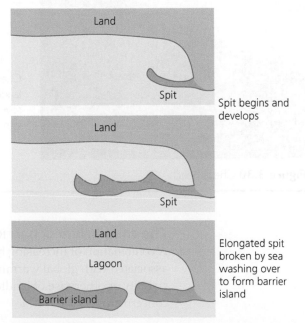

▲ **Figure 3.29** Barrier island formation – spit breaching

LOCATED EXAMPLE: THE ORIGINS OF CHESIL BEACH – AN INTERACTION OF FACTORS

Shingle accumulations around mid-latitude coasts, such as the British Isles, are thought to be more a product of past processes than of those operating today. The vast areas of foreshore exposed when sea level was some 120 metres below where it is today, were subject to intense weathering such as frost shattering. As sea levels began to rise as land-based ice started melting, wave energy picked up this broken-down material. Attrition rounded the sediment and eventually the growing bar of pebbles came to a halt against a cliff line.

A large-scale example is Chesil Beach, Dorset. At Bridport, its western end, Chesil is made up of small pea-sized shingle and is about 4 m high. Thirty kilometres away at its eastern end at Portland, the ridge reaches 15 m high and is made up of 5–7.5 cm pebbles and cobbles. The origins of the sediment are mostly flint from chalk, chert (a silica rock from Greensand), limestones from Portland.

▲ **Figure 3.30** Chesil Beach

Originally thought to have been a spit that grew along the coast to become a tombolo joining the mainland to the Isle of Portland, research now indicates a more complex origin.

- It does seem as if much of the sediment was rolled onshore as sea level rose as the last ice age ended.

- The aspect of the present-day beach is directly to the south-west of where high energy waves arrive. The beach is orientated virtually perpendicular to this direction, suggesting that swash alignment is a factor.

- Longshore transport has a role as the high energy waves from the south-west can move all sizes of sediment. Only the smaller sediment can be moved east to west by the lower energy waves that come from the south-east. Recent research indicates that once wave energy is high enough, the larger sediment travels more rapidly than smaller particles. This is because the pebbles offer a larger surface area for waves to have a purchase on.

- Sediment continues to be added to the system from cliff erosion to the west, for example at West Bay (page 57).

- There is commercial exploitation of the sediment with material being removed under licence; this will affect the results of pebble measurement investigations.

Chesil Beach is a valuable reminder that coastal landforms are often the result of the interaction of several factors and that past processes can be very significant.

The equilibrium of barrier islands

A combination of increasing human pressures on coastal areas and changes associated with global warming such as rising sea level, mean that the stability or otherwise of landforms such as barrier islands is a serious concern around the world.

Some of the threats are:

- sediment eroded from seaward face being transported offshore
- waves washing over the island, carrying sediment from seaward face; material deposited in the lagoon causing barrier to migrate landwards
- some particularly low-lying barriers being completely overwhelmed, with sediment dispersed
- loss of protection from high energy waves for coastline behind the barrier
- loss of barrier islands as places of human activity such as housing, transport and employment such as along the Chandeleur Islands, Louisiana, USA.

Offshore bars

Within the intertidal zone, nearshore bars can develop. They have a variety of shapes and arrangements:

- linear and parallel to the shore
- crescent-shaped and parallel to the shore
- discontinuous
- linear at an angle to the shore.

Their formation is a matter of debate but all theories stress the importance of wave energy. **Breakpoint bars** form as material moves onshore where waves interact with the seabed at depths less than wave base. In addition, material moves offshore, transported by swash. A bar forms where the sediment flows converge. Nearshore bars can migrate both on- and offshore. Onshore movement generally occurs when wave energy is low to moderate. Offshore migration is more often associated with higher energy waves when backwash flows are able to carry sediment along the bed. Typical rates of migration lie in the range 1–10 metres per day, although rates as high as 30 metres per day have been observed. There is also evidence of a cyclical pattern of migration, indicating dynamic equilibrium over a period of several years.

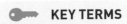

KEY TERMS

Breakpoint bars Accumulations of sediment where waves first break.

Cusps Crescent-shaped features spaced regularly along a beach.

Beach plan: small-scale landforms

Within the overall plan of a beach, a number of smaller-scale features might be present. Their small scale usually results in a short lifespan. Whereas a large beach or bar can persist for decades or centuries, small-scale features such as cusps or ripples may last for only a few days before their shapes are adjusted by changing inputs of sediment and/or energy.

Beach cusps

Cusps range in size from 1 to 60 metres from one point of the crescent to the other. Generally the greater the input from wave energy, the larger the cusp. The horns of the cusp consist of coarse sediment, with finer sediments in the curve of the crescent.

▲ **Figure 3.31** Cusps, Hengistbury Head, Dorset

How cusps form is a matter of uncertainty. However, once developed they reinforce the processes that form them.

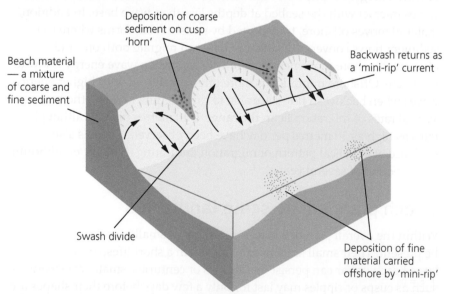

▲ **Figure 3.32** Water and sediment flows in cusps

The model that is best supported by fieldwork evidence combines positive and negative feedback (Figure 3.32). This self-organisation model suggests that along a relatively straight beach face, water is attracted into small depressions and accelerates. Sediment is eroded to make the depression deeper and so it attracts more water and further accelerates the flow as positive feedback operates. Small ridges on the beach repel water and cause flow to decelerate. At these locations lower energy results in deposition, which increases the size of the ridge. As the depressions become larger, negative feedback starts to operate as swash runs out of energy before reaching the back of the cusp and so no sediment is removed. Likewise on the horns of the cusp, once they have reached a certain size water flows off quite quickly so that sediment is not deposited.

Beach ripples

The smallest features in a beach plan are ripples. They form in sand with a scale of up to 10 centimetres in height and up to some 50 centimetres between their crests. Symmetrical ripples form where water flows have similar velocities. Where there is a pronounced flow in one direction asymmetric ripples are found. The strong current tends to encourage the ripple to migrate.

Beach profile

As with any slope system, the beach profile reflects the interaction of a number of variables. Key factors are:

- wave energy
- size and shape of beach material
- tidal range.

The role of wave energy

Both laboratory experiments and fieldwork suggest a strong relationship between wave energy, in particular wave steepness (page 5) and the angle of the beach profile. As wave steepness decreases, beach slope angle increases. It has been observed that the same beach experiences a changing profile on an annual cycle due to variations in wave steepness.

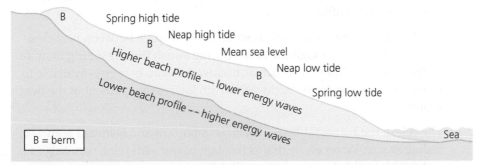

▲ **Figure 3.33** Seasonal changes in beach profile

The more energetic waves of winter erode and transport sediment offshore, perhaps forming an offshore bar. During the summer, lower energy waves move material onshore, building the beach up into a steeper profile. This simplification is fine up to a point, but high energy waves also occur in summer and low energy ones in winter. Terms such as destructive and constructive as applied to waves are not helpful in this context. It is the amount of energy that a wave brings on to a beach that is important. Neither are summer and winter profiles appropriate terms in some regions. Coasts affected by monsoons, which bring high energy waves, often show clear profile changes with the wet and dry seasons, such as around the Indian subcontinent.

The relationship between beach profile and wave energy is complex. Waves moving on to a steep beach tend to break directly on the beach face by surging or plunging. A significant amount of the incoming energy is reflected back from the beach, leading to such beaches being known as reflective beaches. Shallow-angled beaches receive waves that break and spill as wave base is reached further out from the shore. Incoming wave energy is dissipated as the waves move across the wide beach, leading these beaches to be called dissipative beaches.

The role of sediment size and shape

Sediment size

The size or calibre of the slope material is an important influence on its shape and steepness. Steep beaches are associated with larger sediments; shallow profiles with finer sand. The link between sediment size and the gradient lies in contrasting percolation rates.

With coarse-grained sediment, larger spaces between individual particles allow more water to pass through more rapidly. Therefore, as swash moves up a shingle beach, much water drains down through the beach and significant friction occurs between moving water and sediment. As a result backwash is greatly reduced in strength. In terms of sediment transport there is more energy available to move particles up the beach; because the return flow of water is limited, there is little energy available to carry sediment seawards. The overall effect is to build up the beach, thereby steepening the profile.

Smaller calibre sediment, such as sand, has less space between particles. Less water drains through the beach during swash flow, resulting in more water returning down the beach as backwash. More energy is available to transport sediment in the backwash which lowers the gradient of the beach profile.

The role of the water table in a beach is important and is linked with sediment size. Water easily drains through coarse sediment, resulting in a water table well below the surface. This leaves an unsaturated zone into which additional water can drain. Less water flows as backwash on these beaches. Sand, on the other hand, tends to be saturated almost to the surface. Little additional water can drain away and so more water flows across the surface, similar to the conditions that result in overland flow within a drainage basin. More water flowing across a beach surface means

KEY TERMS

Percolation The rate at which water drains through a material.

Water table This represents the upper level of saturation beneath the surface.

more transport of sediment. In particular, a higher ratio of backwash to swash tends to move material offshore, thus lowering beach gradients.

Another factor influencing steepness is particle size: the larger the particle size, the steeper is the angle of rest. Thus pebbles can form a steeper slope than sand; coarse sand steeper than fine sand. Shingle beaches typically develop gradients over 10° while most sand beaches are less than 5°.

All these factors can be seen operating on beaches that store both larger calibre sediment (shingle, pebbles) and sand. There is often a marked break of slope between the steeper shingle/pebble segment and the lower-angled sand segment (Figure 3.34).

Sediment shape
Sediment shape can influence its movement. By using the three axes (Figure 2.10), larger particles such as pebbles can be placed in one of four categories of shape:

- Disc – flat
- Sphere – round
- Rod – long and rounded
- Blade – long and flat.

▲ **Figure 3.34** Contrasting beach slope angles: a sand and pebble beach, Sidmouth, Devon

The relative proportion of the surface area and ability to roll are important factors that influence sediment transport and therefore the distribution of particle shapes across the beach profile.

The role of tidal range

As well as the daily variations in low and high water, tides follow an approximately fourteen-day cycle of high (spring) and low (neap) tides. The difference between the high and low tides is an important influence on beach profiles as it determines the width of beaches. Where coasts receive a combination of high wave energy and a macro-tidal range, beaches tend to be wide. These conditions are found in the British Isles, the west coast of Canada and southern South America (pages 15–6).

On beaches made up of shingle and pebbles, **berms** can be identified. These usually relate to swash processes associated with the levels of high water reached in the tidal cycles (Figure 3.33). At the highest level the berm is often a flat-topped feature and only changes shape due to wave energy when a severe storm occurs at the same time as a spring tide. Sand beaches tend not to have such marked ridges in their profiles. Sand is relatively easily reworked by the wind so a small ridge left at a spring high-water level is unlikely to last intact until the next time the tide reaches this point.

 KEY TERM

Berms Linear beach ridges, usually parallel with the high water mark.

④ Low wave energy coasts – estuaries and deltas

▶ *What forms do estuaries and deltas take and what processes operate in them?*

Estuaries and deltas lack the physical drama of high energy waves breaking against tall and steep cliffs. Nevertheless, they have 'quiet' significance not only for physical processes but increasingly for human activities. Changes to the processes operating in and on them, as well as the landforms themselves, are very significant.

Estuaries and deltas are locations where rivers extend into the coastal zone. Both are important stores of sediment and result from the interaction of marine and fluvial processes. However, there is a fundamental difference between them. Deltas are areas where sediment is accumulating out from the land into the sea. Estuaries are indentations in the coastline, often funnel-shaped, that are infilling

▲ **Figure 3.35** Mawddach estuary, mid-Wales, looking inland

with sediment (Figure 3.35). The storage of weathered and eroded material from land areas in submarine sediments has a significant influence for global climate change. Estuaries are significant sinks for carbon and also contaminants such as heavy metals (e.g. mercury).

Estuaries

Most estuaries can be divided into three compartments (Figure 3.36).

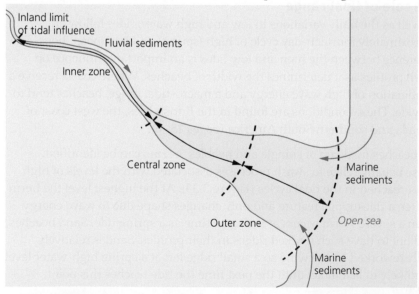

▲ **Figure 3.36** Estuary compartments

In terms of energy inputs, the outer (seaward) compartment receives much wave and tidal current energy, while the inner (landward) compartment has considerable energy inputs from river currents (Figure 3.37).

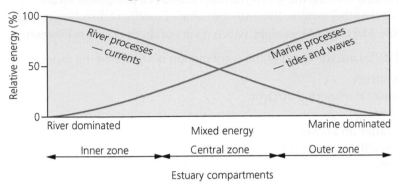

▲ **Figure 3.37** Energy changes within an estuary

Finer sediments are transported through these two areas into the less energetic middle compartment. It is here that most deposition occurs. Coarse material, such as sands and gravel, tends to be deposited in the inner and outer compartments.

The mixing of salt with fresh water is an important but variable process within estuaries. Mixing takes place due to diffusion, a molecular process, when differences in the ionic make-up of salt and fresh water result in turbulence that leads to salinity becoming the same throughout the estuary. Advection is a physical process that mixes salt and fresh water due to the flows of water within the estuary. In any one estuary, both processes can occur at the same time but one often dominates over the other.

Research has shown that how well fresh and salt water mix allows us to identify three types of estuary:

- Stratified – very little mixing
- Partially mixed
- Well-mixed.

Stratified estuaries are found in microtidal environments where neither the tidal nor the river current is strong enough to cause the turbulence necessary to bring about physical mixing. Fresh water lies above salt water as two wedges: fresh water tapers seawards, salt water tapers landwards.

As river and tidal currents increase, the degree of mixing increases. The resulting patterns of fresh, brackish (a mixture of salt and fresh water) and salt water are important influences on ecosystem development.

Although there are significant flows of water into, out of and within estuaries, the dominant process is deposition. Therefore, the term **sediment sinks** is given to estuaries. The stores of sediment, sand and mud can shift location on a frequent basis.

 KEY TERM

Sediment sink Where sediment accumulates and is stored. Different types of sink exist for a wide range of time scales, from days through to millennia.

Deltas

These accumulations of river-derived sediment can be seen at a variety of scales. A small stream entering the sea can develop a delta of a few tens of square metres. At the other extreme there are the deltas of continental-scale rivers such as the Nile, Mekong or Mississippi, which cover hundreds of square kilometres.

Coastal deltas are the result of the combination of the following factors:

● river energy
● sediment transported by river
● wave energy
● tidal range and flows
● shelf width and gradient
● tectonics.

The relative influence of these factors leads to a classification of deltas (Figure 3.38).

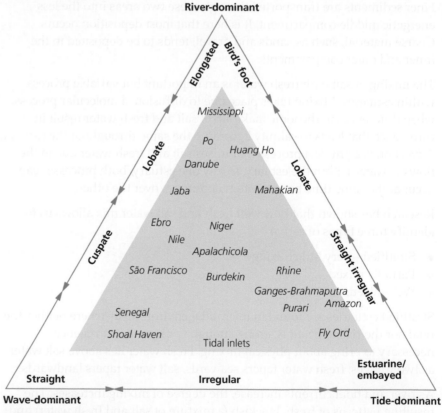

▲ **Figure 3.38** Types of delta

River-dominant deltas are characterised by large catchments, with the river discharging into a relatively low energy sea area. The Mississippi and the Danube are examples. At locations where tidal energy is high and wave energy only moderate, sediment is deposited perpendicular to the coastline. Here tide-dominant deltas form such as the Ganges-Brahmaputra and

Rhine. Deltas that front on to open sea areas and receive high levels of wave energy have smooth coastlines. Strong longshore currents operate and contribute to the formation of linear accumulations of sediment parallel to the coastline. These wave-dominant deltas often have well-developed beaches and sand dune systems. The São Francisco River in the north-east of Brazil has an example of such a delta.

However, it is important to appreciate that a delta is unlikely to owe its shape entirely to one set of processes. Small changes in sea level alter the energy input to deltas as waves and tides extend their influence over more and more of the delta.

Nearly all present-day deltas are relatively young landforms. Sea level has been at its current position for about the last 6000 years, so it is during this time period that deltas have developed, some involving very large volumes of sediment. In the north-west of the Indian subcontinent, two major rivers share in the production of one of the largest deltas in the world. The Ganges and Brahmaputra transport about 1.5 billion tonnes of sediment per year down to the Bay of Bengal. Their delta now consists of some 1500 billion cubic metres of material lying above mean tidal level and nearly 2000 billion cubic metres extending out under the sea.

 # Evaluating the issue

▶ *To what extent are marine cliffs the result of erosion?*

Identifying possible contexts for evaluation

Marine cliffs are the context of this chapter's debate, which seems straightforward enough. However, with any discussion of landforms such as cliffs or beaches, there is a need to reflect on what is meant by 'cliff'. There are two perspectives for a cliff, one its cross-section or profile, the other its plan or bird's-eye view. Some cliffs tower vertically up from the shore while others have a gentle slope angle. There are cliffs that present a relatively straight edge along the coast when seen from above, while others possess indentations.

The issue of 'scale' is an important one in any investigation of landforms. Cliffs vary greatly in terms of physical size. The famous White Cliffs of Dover are made up of chalk which extends some 110 metres from bottom to top, whereas cliffs of just a few metres exist at other locations such as along parts of the Northumberland coast where sand dunes form the back of the beach. The scale of variations in the plan view of cliffs is another consideration, from small indentations of just a few centimetres to larger scale features such as geos and small coves or bays.

Another important aspect to take account of before selecting evidence and examples for analysis is what is meant by erosion.

- Marine erosion is given its energy by waves arriving in the coastal zone. Moving water due to waves breaking as they enter shallow water results in processes such as hydraulic action and abrasion occurring.
- Fluvial erosion can influence marine cliffs, as when a stream takes flowing water over the cliff edge thereby cutting into the cliff top and cascading down across the cliff face.

- Glacial erosion could be happening in the present day at high latitude locations such as the Arctic Circle, for example around the Greenland coast, or areas in South America where glaciers come down to the sea. In other locations past glaciations have left their mark on the landscape where glaciers once flowed, eroding rock to leave steep slopes along the sides of what are now fjords.

The cliff system highlights both the factors and the interactions among these factors that produce marine cliffs. Factors such as geology, weathering, biotic (flora and fauna) and human activities are contexts that are relevant when investigating marine cliffs (Figure 3.4).

The role of erosion

Marine cliffs are simply 'slopes at the seaside' so it is important to consider the processes that operate on slopes. However, marine cliffs have an additional factor that inland slopes do not – the action of the sea. The input of wave energy at the base of a cliff can often be highly significant. Pressures of up to 11 000 kg/m^2 from the impact of the movement and weight of water can erode even the most resistant of rock making up a cliff face. For such forces to be effective, the coast needs to receive very high amounts of wave energy. Some locations do undergo these conditions regularly and frequently, such as the west coasts of Ireland, northern Scotland and parts of the New Zealand's South Island. Other places do not receive high levels of wave energy, such as around the Mediterranean or the northern coast of Australia.

The creation of small-scale features such as a notch or cave is closely linked to marine erosion. The presence of these erosional features increases the susceptibility of the cliff to collapse and thereby sustains a cliff's steep angle.

The role of wave refraction in distributing wave energy along a coastline is significant to variations in the role marine erosion plays along stretches of coastline. By concentrating wave energy, and therefore erosional processes, on headlands, these are locations where dramatic cliffs can often be found. Bays, by contrast, are locations where wave energy is more dispersed along the shore and so erosion is less effective.

It is not just marine erosion that can affect cliffs. Erosion can occur from the action of running water down the slope. This can pick up and transport away soil and sediment from the upper parts of the cliff, contributing to its development. Runoff and surface flows can be significant factors where cliffs are made of impermeable rock types such as granite or clay.

Another aspect of cliff formation and development is the balance between contemporary processes affecting marine cliffs and influences from the past. Landscapes and landforms are rarely just the product of processes happening today. Events in the past can often have a lasting legacy. In this context glacial erosion during the past ice age has had an influence on marine cliffs today. For example, the immensely steep and high cliffs along the side of the Alaskan, Chilean or Norwegian fjords are primarily the result of erosion by ice as a glacier made its way down to the sea.

The role of factors other than erosion

Exploring a cliff as a system naturally leads to considering factors other than erosion.

- **Cliff geology** – both lithology and structure influence marine cliffs as they affect the relative resistance of a particular rock to being worn away. Structure in the form of the angle of dip of rocks can have a particular impact on a cliff's profile. It is also in this context that small-scale variations in a profile can be related to contrasting geology within the same rock face. A bed of more resistant rock will protrude from the cliff face as it is worn away at a slower rate than relatively less resistant strata. Small-scale features such as geos and coves are the product of erosion. However, local geology is a key influence as erosion requires less resistant rocks to have relatively more resistant ones to either side.

- **Tectonics** can be an important influence through processes that bring about folding and faulting of rocks. The angle at which sedimentary strata dip as regards the coast can influence the effectiveness of erosional processes and thereby the development of marine cliffs. Cliffs made up of strata dipping down towards the land tend to offer greater resistance to erosion than those dipping towards the sea, as in the case of the latter rocks can fall down along the sloping cliff face.
- **Sub-aerial processes** – the role of weathering is crucial. A distinction can be made between the upper and lower portions of a cliff's profile in terms of processes. Although wave erosion can often be very effective at the base of the cliff, weathering is not absent in this zone. However, it is perhaps in the upper portion of the cliff that weathering processes come to the fore. When investigating cliff profiles, the role weathering can have becomes a key factor in determining shape.
- **Mass movements** – cliff profiles are either maintained or altered largely by mass movements. However, mass movement is closely linked to the nature of the geology of a cliff as well as the weathering processes acting on it. The role of water within a cliff system is often critical to the type of mass movement. Rocks that can hold water within themselves undergo substantial increases in shear stress due to the weight of water. Additionally, water can act as a lubricant with the cliff thereby reducing shear strength.
- **Changing sea level** will bring about either the submergence or the emergence of a

coastline. Cliff profiles can reflect adjustments such as in the development of slope-over-wall profiles or the removal of a former marine cliff away from erosional influences.

- **The influence of human activities** is an increasingly significant factor in coastal features, and cliffs are no exception. Human activities include removing sediment from the coastal zone which then allows wave energy to impact a previously protected cliff line. Coastal management can also reduce the impact of erosion on cliffs such as the pinning and netting of unstable cliff faces or the protection of the foot of a cliff from wave action by installing a sea wall or rip-rap in the form of large boulders.

Arriving at an evidenced conclusion

There is a temptation to write about erosional processes as if they are a constant feature of environments such as the coast. A helpful evaluative point is to indicate an understanding that, for much of the time, a cliff is relatively 'quiet', that is not that much by way of geomorphological 'work' is going on. Reference to infrequent but highly effective high energy events, such as a storm or even a tsunami, as being responsible for a great deal of marine erosion is useful.

As with all landforms and indeed landscapes, very rarely is just one factor, or in this particular evaluation one set of processes (erosion), responsible for its development. From a geographer's perspective, so often it is the interaction of several factors that determines the morphology of a landform and how it changes. Deciding if one factor is relatively more significant than another is difficult, for past processes continue to influence today's landforms and there are important differences from one place to another.

Chapter summary

✔ Coastlines can be investigated by looking at their plan view. Many coasts are either concordant or discordant in plan view, that is a bird's-eye view, largely due to the influence of geology.

✔ Cliffs are a prominent feature of high energy coasts and show considerable variations in both their cross-section and plan views. The use of a systems approach can be a very helpful framework when investigating cliffs, as it highlights how factors such as geology, hydrology and wave action interact to produce a particular type of cliff. A strong influence on many cliffs is the balance between marine and sub-aerial processes.

✔ As a coastline retreats, various landforms develop. Shore platforms extend seawards from a cliff's base and result from the interaction of several factors.

✔ Low energy coastlines often possess extensive beaches. Beach plans reflect the dominant processes of sediment transport operating in the area and tend to be either drift- or swash-aligned. Spits are a type of beach and can be prominent features of a stretch of coastline.

✔ Cuspate forelands are a distinctive type of sediment accumulation protruding from the coastline, while barrier islands can extend for considerable distances along a coast. Often the equilibrium of a barrier beach system is precarious.

✔ Beaches show considerable variation in their profiles, with wave energy and sediment type (size and shape) being two key factors in their development.

✔ Estuaries and deltas resulting from the interaction of river and marine processes can be prominent along low wave energy coasts. Both estuaries and deltas are important sinks for sediment.

Refresher questions

1 Explain the fundamental differences between concordant and discordant coastal plans.

2 Describe and explain how the relief of the land adjacent to the coastline influences coastal plan.

3 What is meant by the terms 'rock structure' and 'rock lithology'?

4 Describe and explain the influence of rock structure on cliff profiles.

5 Outline how the balance between marine and sub-aerial processes influences cliff profiles.

6 Explain the formation of landforms resulting from cliff retreat.

7 Distinguish between swash- and drift-aligned beaches.

8 Explain the processes operating to form spits.

9 Suggest how barrier beaches might be formed.

10 Describe and explain seasonal variations in the profile of many beaches.

11 Explain the significance of sediment size and shape as influences on beach profiles.

12 Outline why it is helpful to distinguish between different areas of an estuary.

Discussion activities

1 In small groups, study the coastline for stretches of contrasting coast for which you know the geology (rock type and structure) such as the Gower Peninsular (page 52). OS maps at 1:50 000 and 1:25 000 scale would be appropriate. Also the British Geological Survey website has available information concerning rock structure. Discuss how wave refraction affects the distribution of energy along the coastline. You will need to know the direction of prevailing winds and any significant secondary wind directions. How might the distribution of energy influence the landforms you could expect to find along your stretches of coast?

2 Using OS maps or Google Earth, discuss the key factors influencing the development of beaches along a stretch a coastline. What information not given by either a map or image would be useful to aid your discussion?

3 Discuss how and why beaches change their shape through time. Consider such changes across a variety of time scales such as a tidal cycle, seasonally or annually. Include in your discussion locations beyond the British Isles, perhaps making use of experiences among your fellow students of overseas holidays. Sometimes holiday pictures can be a valuable, if somewhat unexpected, resource!

4 Discuss the importance of scale when investigating cliffs or beaches. Consider the range in size of such landforms (for example, cliff heights, width and length of beaches) and how it can be important to appreciate both the smaller and larger-scale features of these landforms.

5 With reference to both high energy cliffed coastlines and low energy coastlines such as sediment accumulations, research how contrasting types of coastal landscapes are represented informally as places. In small groups, choose a particular type of informal representation such as a painting, music, television, film or text (novel/poetry) and investigate how marine features such as beaches, cliffs or estuaries are represented by that medium. Present the findings of your group to the other groups so as to build up knowledge and understanding of the strengths and weaknesses of different informal representations.

FIELDWORK FOCUS

A *Cliffs* These are a difficult proposition to investigate at A-level. There are considerable safety issues to take account of and the type of surveying equipment used is not available to schools. You could examine the landforms along a stretch of coastline or contrasting coastlines in terms of the relationship between features and factors such as geology, aspect and wave energy (e.g. using fetch, average wind conditions). It is possible to estimate cliff height quite accurately using a clinometer and tape measure and simple trigonometry.

Detailed field sketches of contrasting cliff profiles are valuable – perhaps when used as part of an investigation examining the place characteristics of a seaside location. Many coastal locations gain part of their representation from their physical setting – the White Cliffs of Dover and several locations are famous because they have been used as the settings for television programmes, for example. The role of local cliffs in attracting visitors to a location could be an informative element of a survey of visitor attitudes.

B *Beaches* These are much more accessible and feasible as locations for independent investigations, although here again safety issues must be paramount when planning

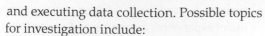
and executing data collection. Possible topics for investigation include:

- Comparing the profiles of beaches in different locations
- Comparing the seasonal profiles of the same beach
- Examining the impacts of management schemes such as groynes/beach replenishment on beach plan and profile
- Examining the influence of longshore drift – comparing accumulations of sediment either side of groynes, for example
- Comparing sediment size and shape along a beach profile
- Comparing sediment size and shape along the length of a beach.

C *Investigating the distribution of pebble shapes across a beach profile.* On a pebble beach there can be differences in the types of pebbles in terms of their shapes across the beach profile. You could carry out such an investigation on one beach providing it is long enough to allow you to compare data from several places. An alternative is to compare a couple of different beaches. Whichever approach you adopt, pebble shape data needs to be collected along transects running across the profile. Another consideration is whether you collect data at intervals all along the transect or compare two sites, one at the seaward end of a transect, the other on the topmost ridge. You need to consider what type of sampling strategy is best suited to the investigation. It is also important to survey the cross-sectional shape of the beach so that you can locate your sampling points accurately.

Further reading

Bird, E. (2016) *Coastal Cliffs: Morphology and Management.* Berlin: Springer

Bridges, E.M. (1998) *Classic Landform Guide: Gower Coast.* Sheffield: Geographical Association

Field Studies Council (2001) *Rocky Shore Name Trail*

Field Studies Council (2003) *Rocks Chart*

Long, A.J., Waller, M.P., Plater, A.J. (2006) 'Coastal resilience and late Holocene tidal inlet history: the evolution of Dungeness foreland and the Romney marsh depositional complex', *Geomorphology*, 82(3–4), pp.309–30

Otvos, E. (2012) 'Coastal barriers – nomenclature, processes and classification issues', *Geomorphology*, 139–40, pp.39–52

Woodroffe, S. (2017) 'Coastal landscapes: processes, systems and change', *Geography Review*, September

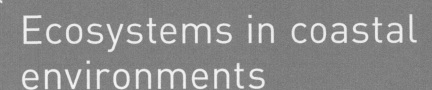

Ecosystems in coastal environments

Previous chapters have focused on non-biological elements of the landforms and landscapes in the coastal zone. On occasions, mention has been made of organisms inhabiting the coast but only as contributors to processes such as weathering and erosion. Increasing attention is being given to gaining knowledge and understanding of the functioning of coastal ecosystems as concerns mount over their protection and conservation, their sustainable use and impacts from human activities and environmental change. This chapter:

- explores the value of coastal ecosystems
- investigates ecological succession and the development of salt marshes and marine dunes
- analyses the formation and characteristics of mangroves and coral reefs
- evaluates the role of vegetation in the development of marine dune systems.

KEY CONCEPTS

Systems A system is a group of related objects. Physical systems tend to be open, having flows of energy and materials across their boundaries. A change in one part of the system brings about changes in other parts through the operation of feedback mechanisms. For example, a mangrove ecosystem can be considered as a system with inputs such as sediment and solar energy, processes and stores such as photosynthesis and nutrients, and outputs such as growth of the mangrove plants and the fish species living in and around the mangroves.

Feedback An automatic internal response to change in a system. For example, in a salt marsh system, if sea level rises more of the marsh will be covered for longer periods of time and with greater depth of water. Plants that can tolerate some submergence but not total immersion in water will die and the marsh will degrade.

Threshold Critical 'tipping points' in a system. For example, with an episode of coral bleaching, if sufficient coral along a reef is affected and dies, the symbiotic relationships between coral and the many species living on and around the reef suffer. If the extent of bleaching is sufficiently great then the entire coral ecosystem degrades and may never recover.

Equilibrium The state of balance in a system achieved when equality exists between inputs and outputs. For example, in a sand dune system, when the amount of fresh sand blown on to the dunes dominated by marram grass is equal to the rate of growth of the marram grass, the dune system tends to remain stable.

1 Recognising the value of coastal ecosystems

▶ *In what ways are coastal ecosystems valuable?*

🔑 KEY TERMS

Ecosystem A functioning, interacting system of organisms and their environment.

Holistic A situation when the parts of something are closely connected to each other and can only be fully understood by reference to the whole.

Biotic Refers to living organisms.

Abiotic The components are non-living, such as geology and climate.

Natural capital The elements of nature that directly or indirectly produce value to people.

Natural income The goods and services that humans gain from natural capital.

Ecosystems are important elements in the physical geography of many, if not all, coasts. Not only can they be actively involved in the development of features such as cliffs and shore platforms, they are also responsible for creating distinctively 'organic' landforms and landscapes such as dune and coral reef systems.

As with any ecosystem, coastal ecosystems are **holistic**. Change in any one part of the ecosystem results in feedback occurring, bringing about change throughout the system. For example, a tropical storm can cause significant coral reef erosion, a fire can destroy dune vegetation, or a cliff collapse can cover an area of shore platform smothering its community.

Plants, animals and micro-organisms occupy habitats in the coastal zone that provide:

- mineral and organic nutrients
- rocks and sediment (to grow on)
- water.

Biotic components are not only influenced by the **abiotic** environment but can also bring about change in their physical environment. Perhaps the most extreme situation is the building of a reef by coral to form an atoll. Plants also play a role in stabilising sediments such as mud and sand in salt marshes and sand dunes.

Ecosystems as assets

The term capital is widely regarded as a factor of production, or more commonly money. Capital is used in various ways to produce goods and services. The term **natural capital** refers to any natural assets that have the capacity to generate goods and services. Natural capital yields **natural income** such as harvests of shellfish or wind energy.

The ways in which the environment is viewed as a supplier of goods has evolved from one of simple exploitation to a position where attention is given to their management in order to achieve sustainability. This does not mean that sustainability is achieved at all times and in all places, but there is recognition of the value of effective management. There is evidence to suggest that some fisheries are being well-managed so that fish stocks achieve a healthy and sustainable level.

As well as tangible goods, increasing attention is being given to the less obvious benefits people obtain from an ecosystem. These societal benefits include clean air, protection from hazards and recreation. The value of shading streets using trees and their role in absorbing pollutants, for example, is well understood and planting schemes have been adopted to realise these benefits. Such **ecosystem services** are now being given prominence in planning and management.

The Millennium Ecosystem Assessment (MEA), established at the start of the current millennium, assessed the consequences of ecosystem change for human well-being. Over the course of four years, just under 1500 scientists around the world appraised the condition of ecosystems as well as identifying changes these ecosystems were experiencing. Arising from the MEA's work was the grouping of ecosystem services into four categories (Table 4.1).

KEY TERM

Ecosystem services
Those natural processes by which the environment provides assets and benefits for human activities.

Ecosystem service type	Definition
Provisioning services	Direct products of ecosystems such as food
Regulating services	Benefits from natural regulation of, for example, CO_2
Cultural services	Non-material benefits obtained from natural systems, such as swimming in the sea or aesthetic pleasure from looking at scenery
Supporting services	Ecosystem processes which support other services, such as nutrient cycling

▲ **Table 4.1** Categories of ecosystem services

The UK Government undertook a National Ecosystem Assessment in 2011 with a follow-up process that surveyed the state of the UK ecosystems. The survey promoted the idea of the value of the ecosystem services that the UK gained, including those in the coastal zone.

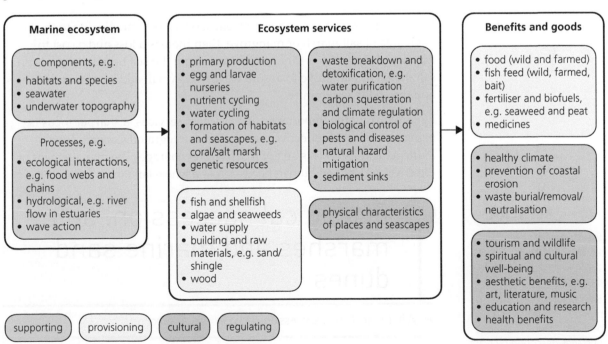

▲ **Figure 4.1** Coastal and marine ecosystem services

Now that more rigorous ways of measuring the value of ecosystems exist, the economic importance of cultural services can be assessed. For example the spending by visitors at four Royal Society for the Protection of Birds (RSPB) seabird reserves was measured in 2009. Visitors to Bempton Cliffs mainly went to watch seabirds, while in the other three locations birdwatching was just one of the reasons people visited (Table 4.2).

Location	Visitor numbers	Estimated spend (£)
Bempton Cliffs, England	76 500	750 000
South Stack, Wales	44 000	223 000
Mull of Galloway, Scotland	19 000	126 000
Rathlin Island, Northern Ireland	14 500	115 000
Totals	154 000	1 214 000

▲ **Table 4.2** Visitor spend at four RSPB seabird reserves in the UK, 2009

In the past, provisioning services have been taken almost for granted. The issues with these services tend to concern sustainability of supply, for example over-fishing can lead to the collapse of fish stocks. The value of considering ecosystem assets more widely, as regulating, cultural and supporting services, is that it allows a broader appreciation of what benefits come from natural capital. These benefits are increasingly being recognised in monetary terms. By doing this, the cost of a reduction in, or even the loss of, a particular service can be calculated. It is then possible to compare costs with benefits which can lead to a more holistic and effective decision-making process.

If it is not possible to use an ecosystem service, for example due to the loss of an area of salt marsh or mangrove, then there is a loss and a cost to humans. An increased threat of flooding from the sea because there is no salt marsh or mangrove to absorb wave energy will add costs to individuals and the wider community.

By recognising the importance of ecosystem services, a sharper focus is being given to their effective management. Integrated Coastal Zone Management (ICZM) is an approach that encourages the management of the environment and human needs across a whole landscape (page 191).

② Ecological succession: salt marshes and marine sand dunes

▶ *What is ecological succession and how does it function?*

The theory of vegetation change within an ecosystem is known as **plant succession**. It involves a number of **seres**, in which each seral community is more complex than the previous one. A succession of vegetation types occurs over time and includes the following changes:

- increasingly favourable physical environment, e.g. soil, water availability, shelter
- progressive increases in nutrient and energy flows
- increases in bio-diversity
- increases in **net primary productivity (NPP)**.

Eventually a **climax community** is reached. So long as the physical environment remains stable, the climax community should persist indefinitely. Because it has greater species diversity, a climax community is seen as being able to sustain itself. It is recognised that a degree of change is most likely and that the term 'mature community' is more suitable. This is intended to describe an ecosystem in which negative feedback prevails, indicating that the community can successfully cope with some degree of stress such as a disease affecting a particular species or a limited time of drought.

Predictable change is, however, rare, and the dynamic nature of ecosystems is seen as important. Three factors are seen as being relevant to change within an ecosystem:

- **autogenic factors**
- **allogenic factors**
- time.

The idea that succession occurs in a regimented progression, following a regular sequence, and is therefore predictable, is a deceptively simple concept. Although changes in vegetation in an area do occur, it is now considered more realistic to think in terms of a dynamic and complex 'mosaic' of plants. Vegetation change can occur for a variety of reasons, not just as a progression from one seral stage to the next. Mature adult plants may be killed, for example by disease or predators, and leave gaps in the ecosystem which opportunistic species might colonise. This may eventually lead to some sort of cyclical change but not necessarily as predicted by succession. Fire, both natural or from human activity (which could be deliberate or accidental), can seriously disturb an ecosystem. In some ecosystems, it is now recognised that fire interrupts succession and that what was once considered as a climax community is in fact a disturbed ecosystem and exists as a subclimax community.

Rather than the emphasis being placed on predictable succession, a more appropriate description of reality is simply of constant transition and of dynamic vegetation change.

Types of coastal ecosystems

Two main types of ecosystems are found in the coastal zone: **hydroseres** and **xeroseres**.

The biotic components found in both types of ecosystem are, to a greater or lesser degree, **halophytic**.

KEY TERMS

Net primary productivity (NPP) The rate at which plants accumulate energy in the form of organic matter, taking into account the energy used in respiration.

Climax community A community seen as existing in a state of equilibrium given the climatic and soil conditions and capable of self-perpetuation.

Autogenic factors Those factors internal to the ecosystem, such as decaying plant material changing the soil conditions.

Allogenic factors Those factors external to the ecosystem, such as climate change or human activities, e.g. trampling pressure.

Hydroseres Communities found in wet or waterlogged conditions such as marsh and mangroves.

Xeroseres Communities found in dry conditions such as sand dunes.

Halophytic Tolerant of relatively high levels of salt in the environment, something that terrestrial plants and animals find toxic.

Coastal ecosystems tend to be prominent where sediment can accumulate. In such locations plants can get a tiny foothold and other elements of food chains and webs can then develop. Superficially a muddy estuary or mangrove-lined coast may not appear to be that significant in ecological terms but it is important to recognise just how productive some of these ecosystems are. In terms of NPP, coastal ecosystems are some of the more productive ecosystems in the world.

Ecosystem	Average NPP g/m²/year
Estuaries including salt marsh	1500
Mangroves	1200
Coral reefs	2000
Tropical rain forest	2200
Cultivated land	650

▲ **Table 4.3** Net primary productivity for selected ecosystems

There is quite a wide dispersion of figures around these mean values depending on local conditions, such as average temperatures and hours of sunlight, but nevertheless, the importance of coastal ecosystems as regards energy flows is clear.

Salt marsh formation

Along many mid-latitude low energy coasts, a type of halosere develops, the salt marsh. In the intertidal zone, factors such as the following impose severe limits on plant colonisation:

- flooding twice daily by tides
- high levels of salt
- wave action
- exposed, windy conditions.

Seagrasses such as eelgrass (*Zostera* spp.) can be found growing as a type of underwater meadow in shallow water just below the low water mark. Pioneer species such as marsh samphire (*Salicornia* spp.) and cord grass (*Spartina* spp.) can tolerate the harsh conditions around the low water level. Plant stems slow water flow which encourages sediment to settle out. Deposition happens when water flow is at its slowest velocity, the lowest part of the tidal cycle and during the periods either side of high tide. Once plants are present, deposition occurs more continuously and at a faster rate and the height of the marsh rises.

▲ **Figure 4.2** Cord grass in early April, Dorset

The process of flocculation

A key physical process in sediment accumulation in estuaries is **flocculation**. The smallest calibre sediments, clays and silts, are transported in suspension, even when water energy is low. A simplistic argument would suggest that they are deposited in low wave energy environments, such as estuaries (page 78), currents that have more than enough energy to keep the very small particles suspended. The period of slack water (low energy), around low and high tide, is typically short and the settling time required for clay and silt sediment is long. Large volumes of water are moved into and out of an estuary. Also there are some estuaries that do receive significant inputs of wave energy.

Clay particles are not simply tiny grains of larger particles such as sand. Clays are formed by the chemical process of weathering. Clay particles possess negative charges that cause the individual particles to stay apart in fresh water. As clay enters an estuary, the positive sodium ions in saline water overcome the repelling negative forces of the clay particles. When individual clay particles come close, they bind together to form larger agglomerations called **flocs**.

As well as this electrochemical flocculation, organic flocculation may take place. At locations where there are many invertebrates, clay particles are ingested by these organisms. The invertebrates, such as crabs and shrimps, digest any organic matter in the clay and then excrete pellets that are essentially clay flocs. These 'organic' flocs are large enough to settle out and be deposited.

> 🔑 **KEY TERM**
>
> **Flocs** Agglomerations formed by the coming together of individual clay particles.

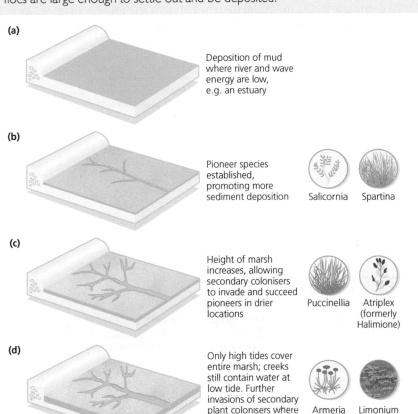

(a) Deposition of mud where river and wave energy are low, e.g. an estuary

(b) Pioneer species established, promoting more sediment deposition — Salicornia, Spartina

(c) Height of marsh increases, allowing secondary colonisers to invade and succeed pioneers in drier locations — Puccinellia, Atriplex (formerly Halimione)

(d) Only high tides cover entire marsh; creeks still contain water at low tide. Further invasions of secondary plant colonisers where conditions are drier and less salty. — Armeria, Limonium

(e) Large areas of marsh remain dry except under very high tide conditions. Ecosystem of upper marsh more terrestrial than marine. — Juncus

▲ **Figure 4.3** Salt marsh succession

▲ **Figure 4.4** Salt marsh at low tide with cord grass the dominant plant type, early April, Dorset

As the height of the marsh climbs up, conditions alter – sediment is drier for longer and less salty. The organic content of the marsh increases due to plant decomposition. Alkaline conditions in the lower marsh give way to more acidic soil further from the intertidal zone. Different plant species are then able to colonise such as sea purslane (*Atriplex* spp.), salt marsh grass (*Puccinellia* spp.), sea lavender (*Limonium* spp.) and thrift (*Armeria* spp.) As the invasion and succession of different species continues, the marsh builds outwards from the shore (Figure 4.3). The area furthest from the open water becomes less marine and more terrestrial in its characteristics. Rushes (*Juncus* spp.) colonise the wetter areas as do alder trees (*Alnus* spp.) until eventually oak (*Quercus* spp.) woodland establishes itself. Land-based organisms colonise or are frequent visitors to parts of the marsh, such as spiders and flying insects, and mammals such as rabbits use the drier marsh. Marine species (crabs and shellfish) dominate in the intertidal marsh. Through the process of invasion and succession, a mature salt marsh has a clear pattern of vegetation across it from open water through to a terrestrial environment (Figure 4.5).

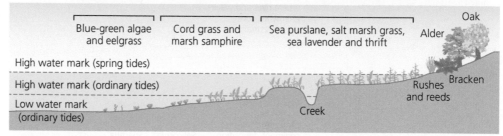

▲ **Figure 4.5** A generalised cross-section across a salt marsh

The precise species succession varies geographically. Different combinations of plants occur around the world.

ANALYSIS AND INTERPRETATION

Study Table 4.4, which shows the rate of annual net accumulation for selected salt marsh locations in England and Wales.

Salt marsh	Rate of annual net accumulation (mm)
1	2.0
2	2.5
3	5.0
4	3.0
5	3.5
6	1.5
7	4.5
8	6.0
9	4.0
10	3.0
11	5.5
12	7.0
13	6.5

▶ **Table 4.4** Rates of salt marsh accumulation for selected locations in England and Wales

(a) Calculate the mean and median rates of net accumulation for the data shown in Table 4.4.

GUIDANCE

There are two statistics to calculate, both measures of central tendency that give a simple overview of the data set. The mean or average requires the data values to be summed and divided by the number of items of data, 13. The median is the middle value of the data set. This is obtained by placing the data in rank order and finding the middle value in the sequence. It is important to work accurately and to set out your working clearly. Although not a high tariff question, it is careless to lose these few marks through being casual. State your answer giving the units the figures are in.

(b) Explain how plant succession forms a salt marsh.

GUIDANCE

It is important to display substantial knowledge and authoritative understanding when using concepts such as plant succession. Opening a response with a clear and accurate definition establishes a sharp focus on the question. The next stage is to securely link the process to the landform/landscape, which in this case is a salt marsh. A key aspect of succession is the invasion and succession of different plant species through time as physical conditions change. Starting with the pioneers, proceed through the sequence of changes both in species and conditions that a salt marsh goes through. Including a few details such as plant names helps make a response more convincing. It might be that a well-labelled diagram conveys much of the knowledge and understanding of the process.

(c) Suggest strengths and weaknesses of viewing a salt marsh as an open system.

GUIDANCE

This question is more open-ended than the previous one and requires an assessment. It is helpful to establish early on that the response is based on a firm understanding of what an open system is and how it operates. A simple diagram might well efficiently achieve this, especially if the labelling used a salt marsh for examples of inputs (e.g. suspended sediment, tides), stores and processes (e.g. accumulation of sediment, plant succession) and outputs (e.g. salt marsh landforms, energy). The crucial aspect is to offer a point of view regarding the advantages and disadvantages of applying the open system idea to a salt marsh. A significant advantage is identifying how change in one component can have effects throughout the salt marsh system. Suggesting possible changes such as rising sea level or reduction in sediment supply and following the effects through the system that such change might have will be convincing. Disadvantages should be considered such as the difficulty of establishing boundaries to open systems. How far inland does the system extend in terms of sediment supply? Should a salt marsh system behind a spit include the spit?

Marine dunes

Coastal sand dunes are common features of many coasts in mid-latitudes. They develop above the high tide level and can extend as much as 10 kilometres inland. Some dune systems consist of a sequence of ridges and troughs parallel to the shoreline. Others have more complex arrangements, and ridges at right angles to the sea or bending away from the beach can be found. The height of the ridges varies from 1–2 metres up to 30 metres. Several factors favour the formation of coastal dunes:

- abundant supply of sand
- low gradient of beach

- strong onshore winds
- area inland of the beach where dunes can develop
- vegetation to colonise dunes.

The interaction between wind speed, vegetation and sand transport

Aeolian transport of sand is mainly by saltation (page 40). On land, this is an energetic process as the falling sand grain has a greater impact than in water. Sand grains can be displaced high above the surface, by as much as 1 metre. Wind velocity increases significantly above the surface, therefore grains can be readily transported downwind. Once sediment is moving, a small increase in wind velocity generates a large increase in sand transport. For example, a twenty-five per cent increase in wind speed brings about a doubling of the sediment transport rate. Grains can also be rolled over a flat surface and slide down a sloping surface.

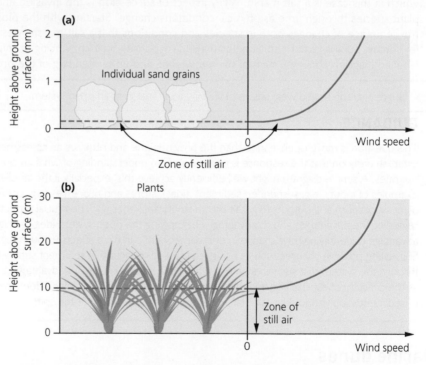

▲ **Figure 4.6** Wind speeds above bare sand (a) and a vegetated surface (b); NB it is important to appreciate the difference in the scales of the *y* axes in the two diagrams

Wind speed increases rapidly with height from just above a bare sand surface. Because of their size (they project above the small zone of still air right at the surface) and rough surfaces, sand grains are particularly susceptible to being moved by the wind.

Where plants are growing, the zone of still air can extend several centimetres above the surface. Plant stems and leaves offer frictional resistance to moving air. This leaves sediment at the surface undisturbed and encourages any moving sediment to be deposited (Figure 4.6).

Formation of marine dunes

Marine sand dune systems are examples of landforms in which plants play crucial roles. The formation, development and stability of dunes depend to a significant extent on plants.

Marine dunes usually begin to form just above the spring high tide level. In the lee of obstacles such as driftwood or seaweed, sand can begin to accumulate due to slightly lower wind speeds. Seeds of pioneer species can germinate here. Their leaves and stems reduce wind speeds while roots also help trap and stabilise sand. Small and low embryo dunes develop. If sufficient sand accumulates, neighbouring embryo dunes merge to give a line of foredunes about 2 metres high that mark the back of the beach.

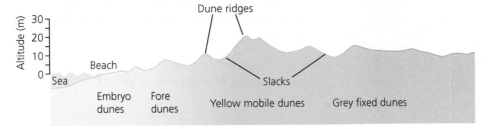

▲ **Figure 4.7** Cross-section across a generalised dune system

As foredunes are colonised by plants, they grow vertically (up to about 10 metres) and in width, forming a substantial ridge. Wind speeds are higher on the windward (seaward) side compared to the lee side. Sand is transported up and over the ridge crest and then deposited on the lee slope. Over time, each dune ridge gradually migrates inland. Rates of movement of up to 7 m yr^{-1} have been observed.

As more sand accumulates, often a sequence of parallel ridges develops that extends the system inland. Between some ridges a **slack** can occur. These are locations where the **water table** intersects the surface. The water table moves vertically depending on the balance between inputs and outputs of water. Along mid-latitude coasts, such as the UK and the Netherlands, the winters tend to be wetter than the summers and so water tables rise and fall seasonally. It is quite possible for slacks to become dry during the summer.

The terms *yellow* and *grey* are given to areas of a dune system, as are *mobile* and *fixed*. These describe changes to the dune system with increasing distance from the sea. Yellow mobile dunes are where sand continues to accumulate on ridges and the top layer of the sand has little organic content so is predominantly a yellow colour. Less blown sand reaches further inland and so the landscape is more subdued. With additional vegetation, a more developed, grey-coloured soil forms containing higher amounts of **humus**. Higher levels of humus mean that more nutrients are available to plants and the capacity of the soil to absorb

 KEY TERMS

Slacks Depressions formed between dune ridges, often containing pools of water.

Water table The upper boundary in rocks between unsaturated and saturated conditions.

Humus The organic matter in the soil derived from the breakdown of plants and animals. It is a major contributor of soil fertility, and increases water retention.

and store fresh water from rainfall is increased. The changing soil conditions allow different plants to become established.

A dune system develops seawards from where it first began. The greater the distance from the sea, the older the dunes are so there is an interesting spatial pattern that also reflects changes through time (Figure 4.7).

Blowouts within a dune system – positive feedback at work

Among the yellow dune ridges examples of blowouts can be found (Figure 4.8).

Their formation is often initiated by the loss of significant amounts of vegetation such as through:

- animal activity, for example rabbit burrows
- wave activity, for example storm waves in combination with a particularly high tide
- human activity, for example trampling of vegetation by walking, horse riding, driving of vehicles.

The wind then removes sand by deflation as the stabilising effect of vegetation is absent.

▲ **Figure 4.8** Dune blowout, Sandscale, Cumbria

(a) Deflation initiated

Wind

(b) Blowout hollow develops

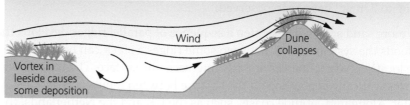

Wind

Dune collapses

Vortex in leeside causes some deposition

(c) Deflation ceases

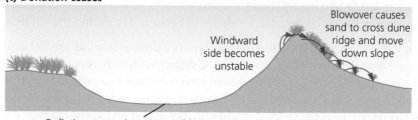

Blowover causes sand to cross dune ridge and move down slope

Windward side becomes unstable

Deflation stops when water table reached or larger-sized sediment encountered

▲ **Figure 4.9** Blowout formation

Once bare sand is exposed, wind flow excavates a deepening hollow, transporting sand inland. As the windward slope steepens, more and more sand is removed by the wind. Vegetation loses its roothold and more sand is left exposed and blown further inland.

Eventually deflation stops. If the water table is reached the sand is too wet to be picked up. Alternatively, it can be that below the sand lies coarser sediment which is too heavy to be picked up by the wind (Figure 4.9).

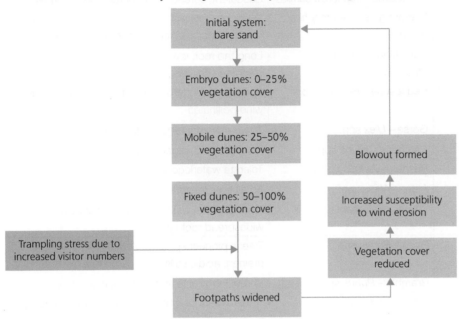

▲ **Figure 4.10** Feedback mechanisms in a dune system

The stress could occur at any location within the dune system. Wherever it takes place, the system alters and, depending on the severity of the disturbance, different feedback mechanisms could operate (Figure 4.10).

The ecology of marine dunes

A beach is a very hostile environments for most plants. Those that manage to become established possess adaptations allowing them to overcome such factors as:

- scarcity of fresh water
- little water retention in the soil
- mobile land surface – sand
- high levels of salt
- exposure to strong and persistent winds
- extreme **diurnal** surface temperatures
- low levels of nutrients.

The pioneer plants that manage to colonise embryo dunes and those that dominate other dune ridges are tough varieties with a range of adaptations. Not only are many species halophytes but they are also **xerophytes**.

 KEY TERMS

Diurnal Something taking place over the course of a 24-hour period.

Xerophytes Plants that are adapted to dry environments.

Location within dune system	Characteristic plants	Adaptations
Embryo dunes	Saltwort – *Salsola kali*	Succulent
	Sea bindweed – *Calystegia soldanella*	Creeps along surface; readily roots from stems in contact with surface
Foredunes	Sand couch – *Agropyron junceiforme*	Spreads by underground rhizomes; narrow leaves; wind pollinated
	Sea spurge – *Euphorbia paralias*	Succulent; spreads by underground rhizomes
Yellow dunes	Marram grass – *Ammophilia arenaria*	Very long roots; spreads by underground rhizomes; narrow leaves curl round; ridged leaf surface
	Dandelion – *Taraxacum officinale*	Long tap root; low-growing leaf rosette; wind-dispersed seeds
Grey dunes	Red fescue – *Festuca rubra*	Spreads by underground rhizomes; narrow leaves; wind pollinated
	Gorse – *Ulex* spp.	Thorn covered; regenerates if burnt
	Heather – *Calluna* spp.	Woody stems; regenerates if burnt
Slacks	Rushes – *Juncus* species	Tolerate waterlogged soil
	Alder – *Alnus* spp.	Small tree; tolerate waterlogged soil
	Willow – *Salix* spp.	Trees and shrubs; tolerate waterlogged soils; widespread root system
Dune scrub	Birch – *Betula* spp.	Tree; a pioneer species tolerant of light, well drained, acidic soils
	Bramble – *Rubus* spp.	Shrub; long roots that produce many shoots
Woodland	Pine – *Pinus* spp.	Tree; tolerant of light, well drained, acidic soils
	Oak – *Quercus* spp.	Tree; climax species

▲ **Table 4.5** Dune plants: their locations and adaptations

KEY TERM

Rhizomes Stems running underground which sprout shoots and roots.

▶ **Figure 4.11** Marram grass about 1.5 metres tall in September, Dawlish Warren, Devon

The respective adaptations that have allowed different plants to colonise and succeed to various locations in the dune system are very significant. The use of the wind for pollination is important, for example. The relative high wind speeds mean that any scent to attract insects would soon be so dispersed that a flower could not act as a homing beacon for the insect. Slack vegetation has to survive extended periods of waterlogging.

Perhaps the key species is marram grass. Its role is crucial to the stabilisation of the yellow dunes and when the supply of sand ceases, marram grass will die. It requires fresh supplies of sand to build up on the surface and will keep growing horizontally and vertically through the sand. Its rhizomes can spread by 2 metres a year.

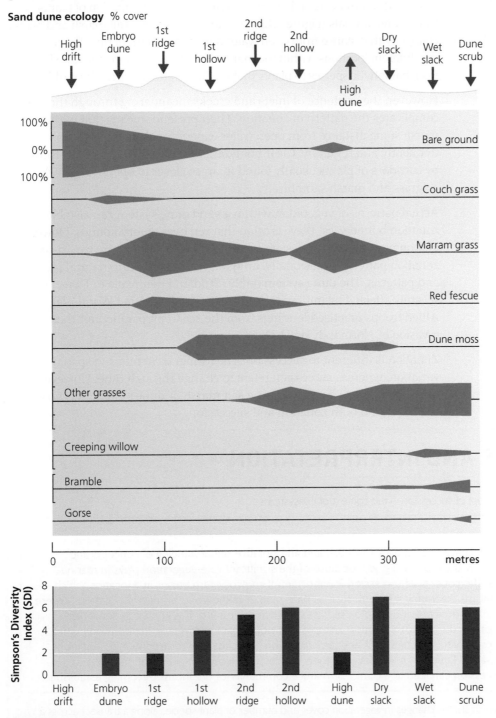

Sand dune ecology % cover

▲ **Figure 4.12** Kite diagram showing variations in vegetation cover and bar graph showing variations in diversity across a sand dune system, Morfa Harlech, North Wales

Zonation exists quite clearly in many marine dune and salt marsh systems. A series of distinct ecological units, each more or less parallel to the shore in the case of dunes, extends inland from the beach. The sequence of dunes, slacks, scrub and woodland each have their characteristic combination of plants, insects and animals (Figure 4.12). The same is true for a salt marsh and plants change with distance from open water due to the length of time the sea covers the different locations within the marsh. In both ecosystems the transition from halophytic species to plants that are intolerant of salt can be striking.

However, the existence of inlets and creeks meandering through the marsh complicates the pattern of zonation. Their presence means that parts of the marsh some distance from open water experience deeper water more frequently and regularly. Clear boundaries to the zones can be cut through by corridors of plants usually found in areas closer to open water such as eelgrass and marsh samphire.

Actual patterns of vegetation within a sand dune system can also blur zonation boundaries. Wave erosion that removes embryo dunes, blowouts and human interventions such as golf courses or agriculture can distort a zonation pattern. Occasionally, unique events can disturb natural processes and patterns. The dune system on the Studland peninsula in Dorset was severely affected by military activities during the Second World War, as Allied troops practised landings from the sea in preparation for their invasion of Normandy on D-Day.

Zonation is different to succession. The former term refers to spatial variation, whereas succession refers to change through time. However, by investigating the changes across a salt marsh or dune system, it is possible to understand both zonation and succession.

ANALYSIS AND INTERPRETATION

(a) Explain the concept of succession in the context of ecosystems.

GUIDANCE

You are expected to have substantial knowledge and authoritative understanding of concepts and processes relevant to any topic you study. Because of the significant role vegetation plays in marine dune ecosystems, the concept of succession is important. If an ecosystem begins at a location which has not had any organisms colonising it before, such as bare sand, it is known as primary succession. Where succession takes place on a previously colonised area, such as after a fire, it is known as secondary succession. Succession describes the gradual changes that occur in an ecosystem. Plants invade when they are able to cope with the conditions and, as succession proceeds, the environmental conditions, such as soil type, alter. Different organisms can then invade and succeed to the location. If left undisturbed, succession can lead to a climax or mature community of organisms becoming established.

Study Figure 4.13, which shows changes in vegetation cover and number of plant species along a transect across a marine dune system.

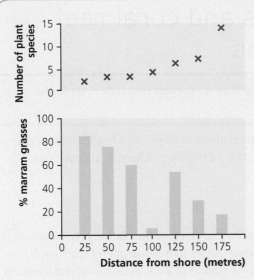

◀ **Figure 4.13** Changes in vegetation cover and number of plant species along a transect across a marine dune system

(b) With reference to Figure 4.13, explain the significance of distance from the shore on the pattern of number of species.

GUIDANCE

When analysing and interpreting any set of data one approach is to offer the overall trend or picture and then focus on any notable details. Offering a summative statement indicates an ability to draw together the data into a coherent perspective. In this example, as distance from the shore increases, so does the number of plant species, a positive relationship. The significance is to be found in the changing environmental conditions the further inland the point on the transect. The sea's direct influence such as salt spray diminishes, wind speeds reduce and various changes to the soil occur. Increased levels of organic matter accumulate in the soil allowing more fresh water to be held and the pH to become less alkaline. All these changes allow a greater diversity of plants to occupy sites.

(c) With reference to Figure 4.13, suggest reasons for the variations in percentage of marram grass cover across the dune system.

GUIDANCE

Marram grass (*Ammophilia arenaria*) is one of, if not the key plant species in a marine dune ecosystem. It is a remarkably tough, secondary colonising plant and allows the dune system proper to start developing. The high percentages towards the front of the dune system can be explained by marram's ability to colonise little more than bare sand. In fact once fresh sand stops blowing on to the plant, marram grass dies out. Its ability to grow strongly upwards and to spread via its rhizomes allows it to thrive in the fore and yellow dunes. The further inland along the transect, the more stable the sand and the more developed the soil becomes. This allows plants other than marram to colonise and to change the conditions even further. The virtual absence of marram grass at 100 metres is likely to be due to this location being a slack. The number of species has not dropped off, suggesting that plants such as alder, willow and rushes have established themselves in this area. The few marram plants are probably surviving on the very driest locations right at the edge of the slack. An alternative interpretation might be that this location is where a blowout is to be found. This could account for the low percentage of marram grass but not the number of species as a blowout is dominated by bare sand where very little vegetation is present.

③ Mangroves and coral reef ecosystems

▶ *What are the contributions made by mangroves and coral reefs to the coastal zone?*

The tropics, with their greater intensity of sunlight and higher temperatures, have the potential to be the location of highly productive and diverse coastal ecosystems.

Mangroves

Mangroves are a group of tree species that can tolerate relatively high levels of salt water. They range from sprawling shrubs to trees capable of reaching 60 metres high. Where there are extensive areas of these trees, mangrove forests or mangals develop. They are restricted spatially to a zone either side of the equator up to about latitude 30° as they do not tolerate frost. Many mangroves have multiple aerial roots which emerge from the trunk above the mud, anchor the tree and assist in the trapping of sediment. The aerial roots also help with oxygen take-up as the waterlogged sediment the trees are rooted in is low in oxygen.

Three major types of location in which mangrove forests occur are:

▲ **Figure 4.14** Mangroves lining a tidal inlet, Okinawa Island, Pacific Ocean

- river-dominated such as deltas, for example Niger delta, West Africa; Mekong delta, South-East Asia
- tide-dominated such as estuaries, for example Northern Territory, Australia
- coral reef islands, for example Grand Cayman, Greater Antilles.

There tends not to be a strict vegetation succession in mangrove forests but different species colonise different types of location within the delta or estuary. Mangroves are a diverse ecosystem with a high level of NPP. As with salt marsh at higher latitudes, mangroves act as nurseries for many fish and invertebrate species. The main contrast with salt marsh is the much greater proportion of biomass above the water level in mangrove forests.

Mangroves act as an effective barrier absorbing wave energy and so protect the shore from erosion. They have withstood storm surges of up to 2 metres and winds of 150 km/hour. However, they are prone to destruction by extreme tropical storms and tsunamis. Where this, and their removal due to human activity, occur an increased hazard risk exists for the coastal zone (pages 132–4).

Coral reefs

Corals are marine **polyps**. Each polyp is a sac-like animal, typically only a few millimetres in diameter and a few centimetres in length. They secrete a protective skeleton around themselves made up of calcium carbonate ($CaCO_3$). Some corals catch their food by using small stinging tentacles. The majority rely for their sustenance on a **symbiotic** relationship with microscopic algae known as zooxanthellae. These algae release nutrients via photosynthesis, which the polyps feed on. In return the algae are sheltered within the hard coral skeleton and obtain some minerals from the coral. The algae contain pigments which give coral its colours.

Cold-water corals

Because of technological advances in deep-water exploration over the past couple of decades, such as manned and remote-controlled submersibles, various discoveries about the biodiversity of the sea have been made. One of the most intriguing has been the revelation that corals are flourishing in deep and cold water environments. At depths of up to 6000 metres, deep-water corals have been seen flourishing in locations all along the coast of north-west Europe (e.g. Darwin Mounds off north-west Scotland), around Australia, New Zealand and Japan and along both the west and east coasts of the USA. There are even corals off the coast of Antarctica. DNA testing to distinguish between species has so far revealed some 3300 species, with the numbers continuing to climb.

They live in the absence of sunlight, so do not have algae, obtaining energy and nutrients by trapping tiny organisms in passing currents. Research has also revealed these coral communities to be very old, with the oldest so far estimated to be 3000–4000 years old. They are a diverse size, ranging from single polyps the size of a rice grain to colonies that reach 10 metres high and some reef communities extending about 40 kilometres along the seabed.

It would be tempting to assume that at the depths they exist at, cold-water corals offer little by way of ecosystem service. The reality is that they are important habitats for a wide range of organisms such as crustacea and fish. In addition to sustaining biodiversity, these corals also act as breeding areas for commercially important species such as ling and various species of the redfish family.

Warm-water corals

These are the corals most people think of, existing across the same latitudes as mangroves, up about 30° either side of the equator. Tropical corals require particular environmental conditions if they are to flourish:

- temperature – mean annual water temperature not less than 18°C, ideally around 26°C
- water depth – 25 metres or less; but corals die if exposed to the air for too long so extend upwards only as far as the low tide level
- salinity – corals require salinity levels >30 000–32 000 ppm
- light – the zooxanthellae algae require light for photosynthesis
- clear water – sediment reduces the light available; it can also clog the feeding tubes of the coral

> ### 🔑 KEY TERMS
>
> **Polyps** Soft-bodied organisms related to sea anemones and jellyfish.
>
> **Symbiosis** The living together of organisms in close association for their mutual benefit.

- wave action – the water needs to be well oxygenated so some stirring up of the water is beneficial but not so much that waves might physically damage the coral.

Types of tropical coral reefs

Four types of reef have been identified (Table 4.6).

KEY TERM

Biodiversity The variety in life in a location. It includes different types of diversity – number of species, genetic diversity within and between these species and the different ecosystems of which they are part.

Reef type	Characteristics
Fringing	Develop from a coastline, often one protected by a barrier island
Barrier	Develop parallel to the shore, some at considerable distance from it. Tend to be large-scale structures, being quite broad and continuous in their linear length
Atoll	Rise from submerged volcanic foundations. Often roughly circular in shape with a lagoon in the middle
Patch	Small-scale reefs that develop landward of barrier reefs or within lagoons

▲ **Table 4.6** Types of coral reefs

▲ **Figure 4.15** A healthy coral reef ecosystem with a high level of biodiversity, Thailand

Various theories have been suggested for the origin of reefs. As there is a variety of types of reef, it is likely that no one theory can be applied to all reefs. As with some other coastal landforms, the role of events in the past, such as sea level changes, are probably significant.

Reefs do not follow a type of succession as occurs in salt marsh and sand dunes, but some reefs do possess a clear pattern of zonation. Different coral species prefer different habitats within a reef structure, with contrasts in wave energy, water temperature and light levels the defining factors. The variation in these types of coral contribute to the great **biodiversity** found on reefs.

CONTEMPORARY CASE STUDY: THE GREAT BARRIER REEF

The vast linear structure made up of coral reefs, **cays**, large islands and lagoons, extends 2300 kilometres along Australia's eastern coast, covering an area of 344 000 km². Because of the Reef's outstanding natural value, it is both a Marine Park designated and protected by Australia and more recently was listed as a World Heritage location by the United Nations Educational, Scientific and Cultural Organization (UNESCO). It is the world's largest collection of coral reefs containing 400 types of coral, 1600 fish species and 4000 varieties of mollusc. Additionally, it is home to many amphibian, bird and mammal species, which rely on the Reef as a habitat, such as the dugong or sea cow and the large green turtle, both of which are threatened with extinction.

The resource diversity represented by the Reef is largely due to its evolution over millennia. It has been exposed and flooded by changes in sea level associated with glacial/interglacial cycles (pages 117–9). During glacials, sea level fell leaving the reefs as flat-topped limestone hills undergoing sub-aerial denudation. Rivers meandered between the hills and the coastline was further east. During interglacials, the rise in sea level created islands and allowed phases of coral growth. Today's Reef has been growing during the past 15 000 years meaning that a unique biodiversity has developed.

Although given some degree of protection by its status as a World Heritage Site and Marine Park, the Reef is a multi-use area with a range of commercial and recreational activities. This brings a complexity to any management of the Reef as there are multiple players (stakeholders) with varying degrees of vested interest in aspects of the Reef.

The Great Barrier Reef Foundation charity was established in response to a UNESCO appeal for citizens to raise money to protect heritage sites

UNESCO has put the Great Barrier Reef on the World Heritage List, inspiring other players to protect it

Global media raise awareness of the need to protect the Reef; the BBC filmed a series about it in 2015

Players (stakeholders) are located in varying **places** and at different **scales**. The **power** to act — and to affect change — is shared among this **network** of **interconnected** players. The most effective changes occur when players work together in **partnership**

Tourist industries and workers put pressure on the government to ensure the Reef is managed sustainably to give long-term environmental, economic and social benefits

Australian universities, including the Institute of Marine Science and James Cook University, research how best to conserve the Reef

The Australian Government pledged to spend £600 million in 2016 to improve water quality around the Reef

▲ **Figure 4.16** The network of players involved in management of the Great Barrier Reef.

Despite the intense focus on the Reef's biological health, it is under constant threat from human activities. Irresponsible tourists and people involved in the tourist industry, commercial fisheries, and agriculture and mining can cause physical damage, over-exploitation and accidental trapping of animals such as turtles and dugong in nets, and excess run-off of fertiliser and sediment which disturb nutrient balance and decrease water quality.

However, threats posed by climate change are perhaps the most serious to arise from human activities. In a catastrophic nine-month heatwave in 2016, 30 per cent of the Reef's corals succumbed to bleaching and died. Due to the interconnections within any ecosystem, impacts of coral death have extended to other creatures such as molluscs and fish, with most suffering decline in numbers as habitat and food disappears. The loss in variety in the Reef is known as 'biotic homogenisation', a tendency towards similarity and a lack of diversity.

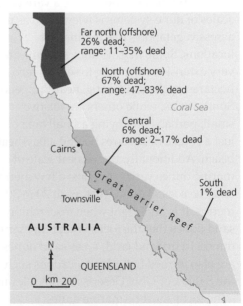

▲ **Figure 4.17** The impact of coral bleaching on the Great Barrier Reef in 2016; the ranges represent upper and lower quartiles

 **KEY TERM**

Cay A small, low-elevation, sandy island on the surface of a coral reef.

4 Evaluating the issue

▶ *Assessing the role of vegetation in the development of marine dune systems over time.*

Identifying possible contexts, data sources and criteria for the assessment

As pressures from human activities in the coastal zone increase, the sustainability of marine dune systems has become an urgent and important issue. Knowledge and understanding of how a dune system operates is, therefore, vital if such ecosystems are to be managed successfully.

- Possible contexts:
  - **(a)** The matter of geographical or *spatial scale* is important to consider in all investigations. In the context of marine dune systems, comparing the development of a variety of scales of dune systems is relevant so as to assess vegetation's role in different locations. Some marine dune systems are very extensive extending to several tens of hectares, such as at Kenfig, near Bridgend, South Wales, while others are relative small, perhaps occupying a small area of just a couple of hectares behind a bay-head beach. Additionally, the vertical scale of marine dunes varies from just a few metres to some coastal dunes that reach 20 to 30 metres. It is important when researching sand dunes that the focus is kept on marine dunes. In dryland (arid) areas sand dunes are also formed but inland locations such as the Sahara or Gobi Deserts are not relevant – the focus must be marine dunes.
  - **(b)** As well as spatial scale, temporal scale can be considered. The macro or geological time-scale is not appropriate when investigating a present-day ecosystem, but

the meso, thousands of years, or micro, a tidal sequence or single storm, may well offer helpful perspectives. Comparing and contrasting dune formation at different scales will offer opportunities to put forward convincing and comprehensive assessment.

- Data sources – a key possible source in the context of marine dunes might well be a student-led fieldwork investigation. Other sources of data on actual sand dune systems can be found among the websites of local wildlife trusts, nature conservation groups and a government agency such as the Environment Agency. Sites run by reputable academic institutions such as those with .ac in their address can contain useful material. However, not all factors are likely to be capable of being quantified, such as volumes and flows of sediment. Wind speed and direction can be measured but require annual averages for the influence of wind to be assessed. Additionally, the issue of extreme atmospheric events, such as a storm, can be significant. Data accumulated over time can be available but will not extend more than a few decades at most for the vast majority of locations.
- Criteria for assessment – as with data sources, some of the criteria are difficult if not impossible to gain information on. However, this does not mean that their possible roles are to be avoided so that informed judgements should be made concerning such points as the availability of sediment. Quantifiable evidence of various possible influences of some factors is available, such as vegetation density.

Assessing the role of vegetation

The development of dune systems relies on sand becoming stabilised, something that starts and continues with plants. Once pioneer species such as sand couch grass become established, sand accumulation can start (page 96). Wind speeds close to the ground are reduced meaning that sand grains are less likely to be transported. The roots of vegetation also help bind the loose soil.

As long as negative feedback occurs, that is more sand accumulates than is lost from the location, then the dunes can develop. Different plants can invade and succeed, such as marram grass and dandelions on the yellow dunes, red fescue on the grey dunes, and alder and rushes in the slacks.

Plants also help transform the sand into soil. As plants die, the decomposition of their tissues adds organic matter to the surface layer. The more humus in the soil, the more the process of vegetation succession can proceed.

Vegetation creates habitats for animals and insects such as spiders, caterpillars and beetles. In turn, insects become the prey of birds, insects and mammals and so food chains and webs develop which are important aspects of a mature dune system.

At locations where human activities remove vegetation, such as where trampling from horse riding or walking are at high levels, the importance of plants can be seen. The formation of blowouts can all too easily lead to serious erosion of a dune system as positive feedback operates.

Assessing the role of sediment supply

Of course, vegetation is not the only influence on dune development. Without a plentiful supply of dry sand, accumulation of sand into a dune system cannot take place. A rocky coastline with cliffs and deep water close to land does not allow dunes to develop. Sediment in the coastal zone is mostly supplied from inland areas, carried down to the coast by rivers. If this supply is limited, perhaps due to low rates of weathering and erosion, then the shortage of sediment will have a negative impact on dune formation and development (pages 36–45). Sand beaches are needed for dunes to develop.

Sand can become mobile as the tide falls and the shore area dries out. The wider this shore area, potentially the more sand is available. A wide shore area is often found along coasts that experience macro-tidal conditions, that is a large tidal range. The west coast of Europe such as Brittany and Cornwall are examples of such locations.

Some plants require fresh sand to blow on top of them, such as marram grass. The grass then grows upwards, extending its roots and rhizomes and keeping on growing. Once the supply of sand stops marram grass dies out, an example of positive feedback.

Assessing the role of wind energy

Factors such as wind speed and direction are important for the transport of sand grains. If there are too few times when strong onshore winds blow, then dune development will be held back. Sand needs to be blown inland allowing it to be accumulated, in particular by marram grass, and for a sequence of ridges and hollows to form. Sand blown offshore is then 'lost' as it lands in the sea and settles out on the seabed.

Assessing the role of time

The research that gave rise to the concept of succession was based on observations of changes in vegetation through time. Sequences of plant communities follow one another, emphasising the role of time in the development of an ecosystem. Very often, investigating a transect across a marine dune system is seen as representing a 'journey through time', with the youngest plant community located on the embryo dunes and the oldest in the woodland.

The rise in sea level following the end of the last major glacial period means that the current shoreline around many coasts has been relatively stable for about the past 6000 years. Marine dunes are, therefore, relatively recent landforms in geomorphological terms. They are also capable of rapid change, for example the impact of a single high energy storm can erode much of a dune system due to wave and wind energy and thereby impact dune development. High energy waves can quickly remove embryo dunes resulting in the dune system apparently starting with the smaller scale yellow dunes at the back of the beach.

Assessing the role of human activities

The growing pressures many coastal dune systems are under from human activities is increasingly apparent around the world. Sand dune locations are used for recreation, housing, industry, transport, forestry, military and agricultural purposes. Sometimes a dune system is completely eradicated but in many locations disturbance is caused that arrests natural changes in vegetation development. An excess of recreational activities can trample or erode vegetation so that only plants resistant to such pressure can survive, or can even lead to a complete loss of vegetation from an area. Fire resulting from human activities can destroy vegetation and seriously reduce biodiversity. When this occurs, different plant communities can establish, for example because the soil is already more 'mature' than bare sand due to the organic matter accumulated before the fire occurred.

Human activities are not all negative as regards ecosystems. Management for conservation is allowing dune systems both the time and space to function without destructive human activities

taking place. In some locations, an absence of disturbance is a side effect of the human activity. For example, military activities or industrial installations often exclude people from large areas. Many coastal steelworks extend across hundreds of hectares of coastal land, much of which is not actually used for the industrial buildings. In the more remote areas of the site, ecosystems can flourish.

Arriving at an evidenced conclusion

Clearly vegetation has a crucial role to play in marine dune formation. Plants begin the process of sand accumulation by physically trapping sand among stems and roots. Plants also provide organic matter which is needed for soil to form. If plant succession operates without major interruptions, then eventually the dunes become 'fixed', vegetation covers the surface and a climax woodland community becomes established.

On the other hand, factors such as sand supply and onshore winds play important roles in dune development. As with all other systems, the flows of energy and materials help characterise and define the relationships among the various components of the system. A case can be made for the availability of sand to play a very significant role, as without this the key species of marram grass does not survive. Likewise, without wind energy, sand is not mobile and a fundamental characteristic of a marine dune system is missing.

Increasingly human activities can either promote the sustainable development of dune systems or lead to the damage and possible destruction of dunes. As with all natural systems, the impacts of anthropogenic global warming are only just beginning to be seen, which in the case of marine dunes is the rise in sea level.

Chapter summary

✔ Various ecosystems, either haloseres or xeroseres, are important components in many coastal locations. The 'ecosystem services' they offer are increasingly being recognised and valued so that their value can be taken into account when managing the coastal zone.

✔ Ecological succession can take place through a sequence of seres at any one location, allowing investigation of how and why an ecosystem varies spatially and temporally. As a location is colonised and transformed, flows of nutrients and energy become more complex.

✔ Salt marsh and sand dune systems are two common and highly dynamic ecosystems located within the coastal system. Both these ecosystems have sets of characteristic plants that require specific adaptations in order to survive in the relatively harsh conditions in both sand dune and salt marsh environments.

✔ Mangroves and coral reefs are two types of tropical coastal ecosystems with high levels of biodiversity capable of thriving in the coastal zone.

✔ All coastal ecosystems are vulnerable to various threats such as rising sea level, increasing water temperature and a range of human activities. The latter, however, are also capable of conserving environments and ecosystems.

Refresher questions

1 Explain the ways ecosystems can be valued as services.

2 What is meant by the terms: 'hydrosere' and 'xerosere'?

3 Outline the process of flocculation.

4 Explain how plant succession operates within either a salt marsh or a marine dune system.

5 Explain why marram grass is so successful at colonising mobile dunes.

6 Describe and explain how a blowout in a dune system can form.

7 Outline the ecosystem services mangroves can provide.

8 Describe the environmental conditions warm water coral requires to flourish.

9 Outline the differences between 'succession' and 'zonation'.

10 Suggest how positive and negative feedback can operate within any one coastal ecosystem.

Discussion activities

1 In small groups explore the advantages of viewing ecosystems as offering services providing goods and benefits. Consider whether giving a value to ecosystem services might encourage their intensive exploitation to realise this value.

2 Discuss the ways in which abiotic conditions influence how ecological succession proceeds in salt marshes and marine dunes.

3 Outline how a coral reef operates as an open system (page 2).

4 Discuss the role that the transport of sediment plays in the formation and development of salt marsh, sand dunes and mangroves. You might consider the ways in which sediment is accumulated in each of the ecosystems, emphasising the role of vegetation in the process of sediment accumulation. Refer back to the material on estuaries and water flow within these, in particular in Chapter 3, to help inform your discussion.

5 Discuss the role coastal ecosystems can play in sediment accumulation in the context of the carbon cycle. Focus on the significance of locations such as salt marsh and mangroves acting as carbon sinks and consider the implications of rising sea level for this role.

FIELDWORK FOCUS

Both salt marsh and marine sand dunes offer a variety of opportunities for the A-level independent investigation. As with all investigations, safety issues must be considered, in particular for coastal investigations the time of tides and speed at which the tide rises. Deep mud and localised quicksands can also be hazardous.

A *Investigating changes in vegetation along a transect.* This can be carried out on both dunes and salt marsh, which allows you to investigate succession. Diversity of plants, their heights and adaptations can be investigated. Along a dune transect in particular, changes in factors such as soil depth, organic content, moisture levels, pH values and wind speed can be usefully investigated and related to changes in vegetation.

It is best to place these changes in the context of the changing profile across the dunes.

This is less easily achieved on a salt marsh as very often changes in slope angles are slight and difficult to measure using the equipment most likely available at A-level.

If the dune complex is managed then comparisons might be possible between managed and unmanaged areas, either on the same dune system or between two different dune locations.

B *Investigating the human geography (in particular, place meanings and representations) of dunes or marshes.* There are interesting synoptic geography themes which you might study independently. For instance, as with other coastal features, both dunes and salt marsh can be very influential factors on place characteristics. Informal representations of both exist in art, literature and music and could be investigated using fieldwork out in the environment. Surveys of the perceptions of visitors and locals are another possible area for investigation in the context of dunes and marshes.

Further reading

Aagaard, T., Orford, J., Murray, A.S. (2007) 'Environmental controls on coastal dune formation; Skallingen Spit, Denmark', *Geomorphology*, 83(1–2), pp.29–47

Barbier, E.B., Hacker, S.D., Kennedy, C., Koch, E.W., Stier, A.C., Silliman, B.R. (2011) 'The value of estuarine and coastal ecosystem services', *Ecological Monographs*, 81(2), pp.169–93

Bridges E.M. (1998) *Classic Landform Guide: North Norfolk Coast.* Sheffield: Geographical Association

Field Studies Council (1997) *Sand Dune Plants Identification Chart*

Field Studies Council (1999) *Saltmarsh Plants Identification Chart*

Hogarth, P. (2007) *The Biology of Mangroves and Sea Grasses.* Oxford: Oxford University Press

Hesp, P. (2002) 'Foredunes and blowouts: initiation, geomorphology and dynamics', *Geomorphology*, 48(1–3), pp.245–68

Lee, S.Y., Primavera, J.H., Dahdouh-Guebas, F., McKee, K., Bosire, J.O., Cannicci, S., Diele, K., Fromard, F., Koedam, N., Marchand, C., Mendelssohn, I., Mukherjee, N., Record, S. (2014) 'Ecological role and services of tropical mangrove ecosystems: a reassessment', *Global Ecology and Biogeography*, 23(7), pp.726–43

Middleton, N. (2008) *The Global Casino* (4th edition), Chapter 7, London: Hodder Education

Perillo, G., Wolanski, E., Cahoon, D., Hopkinson, C. (2018) *Coastal Wetlands: An Integrated Ecosystem Approach* (2nd edition). Oxford: Elsevier

Skinner, M., Abbiss, P., Banks, P., Fyfe, H., Whittaker, I. (2016) *AQA A level Geography* (4th edition) Chapter 6, London: Hodder Education

Coastal system dynamics – changing sea level

The causes of sea level change over very different time scales are varied and include the pressing contemporary issue of rising sea level associated with global warming. Communities around the world are facing challenges resulting from the implications of rising sea level.

Because sea level has risen and fallen, the landscapes and landforms resulting from these changes are different. The consequences of sea level change long ago continue to influence the morphology of the coastal zone, indicating the need to acknowledge the role of past processes as well as present-day ones. This chapter:

- investigates the causes of sea-level change
- explores the landscapes and landforms associated with a relative fall or rise in sea level
- analyses the risks and management of coastal flooding
- assesses the role sea level change can have on coastal landforms.

KEY CONCEPTS

Systems Groups of related components. In physical geography, they tend to be 'open', that is having both inputs and outputs. Sea level is one of the components and as this changes, either rising or falling, the processes and landforms of the coastal zone alter. For example, a fall in sea level results in some landforms created by marine processes no longer experiencing these and sub-aerial processes becoming more significant.

Equilibrium The state of balance within a system. Change in sea level impacts equilibrium, both of long stretches of coastline and also at the smaller scale of individual landforms. As sea level rose as the last ice age began to end, change in sea level was relatively gradual allowing adjustments in the coastal zone to occur. However, the recent rise in sea level due to global warming is beginning to cause increasing instability within the coastal zone, such as increased erosion of cliffs at their base.

Feedback An automatic response to change within a system. For example, as relative sea level falls a new cliff line can emerge which is subject to marine erosion. An increase in volume of sediment could result, which leads to an increase in the effectiveness of abrasion on the cliffs. Further wearing away generates yet more sediment and so positive feedback occurs.

Threshold Critical 'tipping points' in a system. For example, as sea level rises due to global warming, increasing water depth covering coastal ecosystems such as salt marsh, mangroves or corals could result in severe disruption to the functioning of ecosystems so that they degrade permanently.

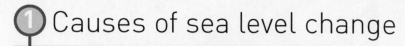

① Causes of sea level change

▶ *What are the causes of sea level change?*

Sea level change is an important factor not just for the coastal zone but for the land as well. The height of sea level determines the base level for erosion by rivers. Rivers cannot erode nor transport below sea level and their gradients influence the energy they possess. Relative movements of land and sea can also significantly alter the area of land exposed to sub-aerial processes. Over the past 100 **Ma**, the relative proportion of land and sea has changed substantially.

Relative sea level change

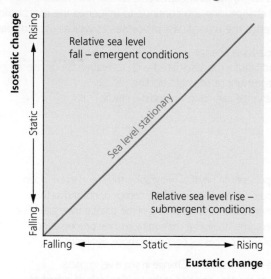

▲ **Figure 5.1** The relationship between isostatic and eustatic changes and relative sea level changes

In many cases, it is not possible to be certain about the precise cause of an observed change in sea level. What really matters is **relative sea level change**. A relative *rise* in sea level can result from:

- sea level rising while the land subsides, remains stationary or rises at a slower rate than the sea
- sea level remaining stationary while the land subsides
- sea level falling while the land subsides at a faster rate.

A relative *fall* in sea level can result from:

- sea level falling while the land surface rises, remains stationary or subsides at a slower rate than the sea
- sea level remaining stationary while the land rises
- sea level rising while the land rises at a faster rate.

A positive sea level change gives rise to transgressive conditions. These lead to a drowning of coastal areas and/or the migration onshore of some landforms such as beaches. Regressive conditions result from falling sea level. Emergent landforms develop as the coastline builds out from its previous position such as abandoned cliffs.

Relative sea level changes can occur on a variety of spatial scales. Global scale result from eustatic changes, while regional- and local-scale adjustments are the result of either uplift or subsidence of the land.

Eustatic change

Changes in absolute seawater level are called eustatic changes. They are global because all the oceans and seas are interconnected. The total volume of global water – liquid water, ice and atmospheric water vapour – is constant.

Component	Surface area (million km²)	Volume of water (thousand km³)	%
Oceans and seas	361	1 370 000	93
Terrestrial waters (e.g. groundwater/lakes/rivers)	134	64 000	5
Land-based ice	16	24 000	2
Atmosphere	510	13	0.001

▲ **Table 5.1** Estimated volumes of the main components of the hydrosphere

It is helpful to think of the hydrosphere as a system, so that a change in the volume of water in one component means that there are changes in other components. In this context it is important to consider the average length of time water stays in the various sub-parts of the hydrosphere.

Changes in where and how water is stored cause eustatic change. Of all the stores, land-based ice is the most significant to eustatic change. For example, if all the land-based ice were to melt, sea level would rise about 90 metres, of which Antarctica would contribute 57 metres and Greenland 7 metres. If all the atmospheric water were to fall from the skies, sea level would rise by some 36 millimetres.

Locally, atmospheric water can make a difference in the short term. For example, the Bay of Bengal can rise by up to a metre during the monsoon season when vast volumes of rainfall land on the Indian subcontinent, travel down rivers such as the Ganges and enter the sea. This additional water then disperses through the oceans and the bay returns to its long-term level.

KEY TERMS

Transgressions The advance of the sea across former land areas.

Regressions The retreat of the sea from the coastline, exposing more land.

Eustatic Refers to worldwide change in sea level.

The hydrosphere Includes all the Earth's water, both fresh and saline, in any state: liquid, solid or gas.

Water store	Average time in store
Atmosphere	10 days
Rivers	14 days
Lakes	10 years
Polar ice	15 000 years
Oceans	3600 years

▲ **Table 5.2** Average time water stays in selected stores

Isostatic change

KEY TERM

Isostatic Refers to movements of the land. These can be upwards if weight is removed, for example by denudation, or downwards if weight is added, for example by ice accumulating.

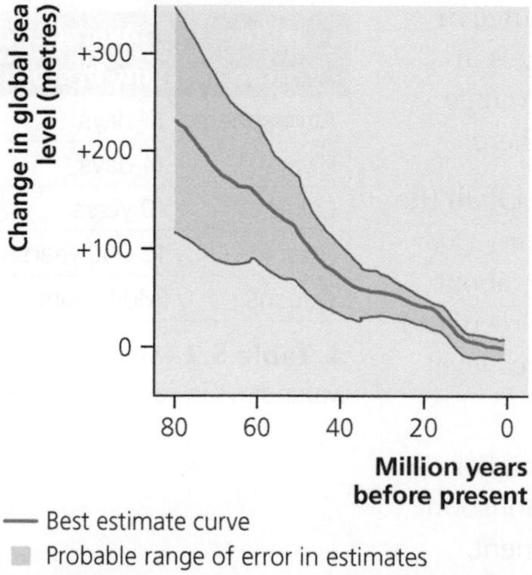

Changes in the absolute level of the land are called isostatic changes and are localised rather than global, as eustatic are. The Earth's crust is capable of being depressed into the semi-molten upper mantle when great weight is added, such as accumulation of ice or sediment. The Mississippi delta has sunk by some 165 metres in the past 10 000 years due to deposition of sediments transported by the river from its basin.

Smaller-scale sinking can result from human activity. Abstraction of water, oil and gas from underlying rocks can cause widespread subsidence. In the Tokyo region, a reduction in ground level of about 4.5 metres has occurred due to groundwater abstraction.

At longer time scales and over larger areas, deposition of sediments in ocean basins depresses oceanic crust. Hand-in-hand with this is the reduction in volume of the basin as sediments build up. Add to this the possibilities of sediment removal from the seabed, either through tectonic uplift out of the water or by subduction at a destructive plate margin, and you begin to appreciate how complicated sea level change is to plot.

Tectonic activity can cause land to rise or fall. Movement along a fault can lift land by several metres in a sudden event. Along both the Alaskan and Californian coastlines, lengths of coast have been elevated by tectonic forces. On a larger scale, the collision of India and Eurasia formed the Himalayas and the Tibetan Plateau. The continental crust in this region was thickened and the continental area reduced. With more ocean basin volume to fill, it has been estimated that sea level fell by about 18 metres.

Ocean basins change shape over geological time. In places, ocean floors sink, increasing the capacity of the ocean to store water. Seafloor spreading widens oceans such as the Atlantic, increasing their capacity. The growth of mid-ocean ridge systems has a significant influence on the volume of ocean basins. Present-day ridge systems occupy a volume equivalent to about 12 per cent of the total volume of the oceans (Figure 5.2).

Exploration by energy companies has yielded much geological information that has increased our knowledge of sea level changes. Details over these very long time scales are patchy, but the overall trend is broadly established from the material recovered from drilling for hydrocarbons (Figure 5.3).

— Best estimate curve

▪ Probable range of error in estimates

▲ **Figure 5.2** Change in global sea level due to changes in volume of mid-oceanic ridges

Pearson Edexcel

AQA

OCR

WJEC/Eduqas

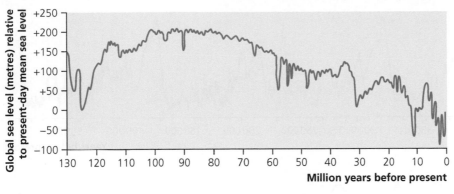

▲ **Figure 5.3** Change in sea level during the past 130 million years

Evidence from marine sediments, such as their lithological and organic characteristics, can help indicate the depth at which the material was first laid down. In addition, it is possible to plot the extent of rocks formed under marine conditions and this gives an approximate trace of former shorelines and when and where these have changed. The use of seismic techniques to reveal the underground geology has also contributed to our knowledge and understanding of very long-term sea level changes.

Sea level changes associated with glaciation

The most recent geological period, the Quaternary, began about 2 million years ago. Geological periods are divided into epochs, of which the Quaternary has two, the Pleistocene followed by the Holocene.

Geological period	Geological epoch	Approximate year began (before present)
Quaternary	Holocene	11 700
	Pleistocene	2.6 million
Tertiary	Pliocene	5.3 million
	Miocene	23 million

▲ **Table 5.3** Recent geological time scale

In the context of studying the present-day coastal zone, it is important to be aware of but not distracted by variations in geological dates. For example, some researchers suggest the Pleistocene began about 1.6 million years ago, others go for closer to 3 million. Differences arise because of the diversity in the types of data, their interpretation and geographical location. In general the end of the Pleistocene receives greater agreement.

What is clear is that the Pleistocene was a period of climatic instability after some 50 million years of very slow cooling with some small-scale fluctuations. Alternate **glacials** and **interglacials** occurred with the consequences for the coastal zone of periods of falling and rising sea levels (Figure 5.4).

 KEY TERMS

Glacial When ice accumulated and advanced over the Earth's surface.

Interglacial Period of warmth between glacials, when ice sheets and glaciers retreated.

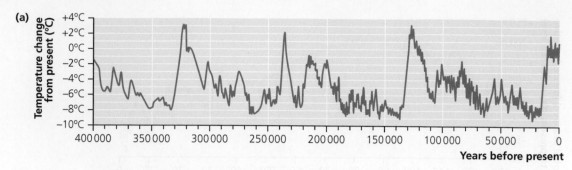

(a)

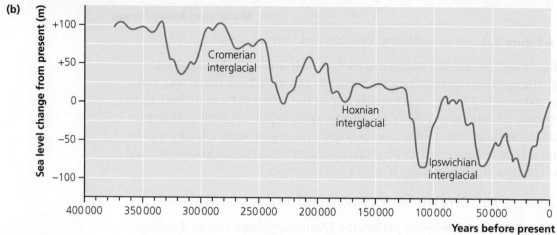

(b)

▲ **Figure 5.4** Temperature and sea level fluctuations during the past 400 000 years

The last major ice advance peaked at around 18 000 years BP. Huge ice sheets extended across vast swathes of the northern hemisphere, covering both land and marine areas. For example, the Scandinavian ice sheet covered the Baltic, Irish and North Seas (Figure 5.5).

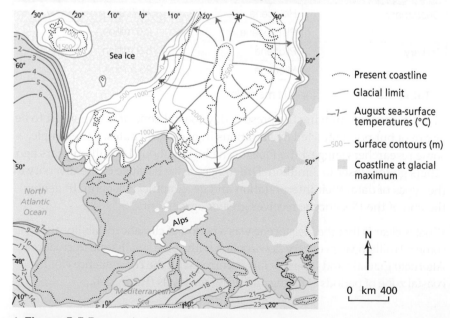

▲ **Figure 5.5** Europe during the last glacial maximum

When land-based ice melts, the freed water flows back to the coast, causing sea level to rise, a process known as **glacio-eustasy**.

There is agreement about the general pattern of sea level change since the last glacial but details are disputed.

Pearson Edexcel

AQA

OCR

WJEC/Eduqas

> **KEY TERMS**
>
> **Glacio-eustasy** Oscillations in sea level caused by the advance and retreat of ice sheets and glaciers.
>
> **Glacio-isostasy** The sinking down of land due to the weight of ice sheets and the upward movement of land when ice melts and weight is lifted off.

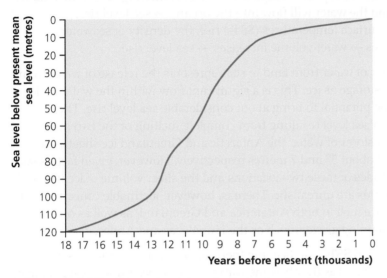

▲ **Figure 5.6** Changing sea level since the last extensive glaciation

It is important to remember that the addition to, or melting of, ice floating in the sea has no effect on the eustatic sea level. The weight of this ice is already supported by the water and so is contributing to sea level.

As ice builds up on land, crustal depression occurs; when the ice melts, weight is removed from the land and the land adjusts upwards, a process known as **glacio-isostasy**.

In general, glacio-eustasy and glacio-isostasy occur at different rates. Glacio-eustasy is relatively rapid in geological terms, whereas time-lags involved with glacio-isostasy (both rising and falling) are long. There is academic debate as to ice sheet thickness, but rates of post-glacial uplift in the Baltic region indicate a rise of some 300 metres having taken place over the past 10 000 years. Isostatic adjustments are continuing as the land continues to 'rebound' at about 10 mm yr^{-1}. Across northern Britain ice sheets were not as thick, with the result that rates of residual uplift are lower, up to 2 mm yr^{-1}.

In a study of coastal landforms and landscapes it is important to appreciate that as positive or negative adjustments in relative sea level occur, energy inputs to the coastal zone change. For example, as sea level falls during a glacial, the intertidal range falls lower and lower leaving parts of the coastline no longer under the influence of maritime processes. Sub-aerial processes begin to dominate. However, as an interglacial proceeds and sea level rises, marine processes once again act on former shorelines.

The influence of global warming on sea level

Global warming, resulting from an enhanced greenhouse effect, is currently responsible for some eustatic change:

1 If average atmospheric temperatures rise sufficiently, ice on the land will melt and the water will flow into the oceans → sea level rise.
2 As sea surface temperatures (SSTs) rise, the density of seawater decreases → water volume increases → sea level rise.

The transfer of water from land to sea represents the release of water from long-term storage as ice. This is a significant flow within the water cycle and has the potential to bring about considerable sea level rise. The increases in sea level resulting from complete melting of the two main land-based stores of water, the Antarctic and Greenland ice sheets, are dramatic at about 57 and 7 metres respectively. However, given factors such as the latitudes of these two locations and the sheer volume of ice involved, such scenarios are unrealistic. There is, however, justifiable concern about the rate of ice melt in both Antarctica and Greenland, as well as other places containing glaciers such as the Himalayas and Andes.

The increase in SSTs results in the thermal expansion of water, as water molecules expand as they become warmer and so occupy a greater volume. It is this process that has contributed the majority of sea level rise due to global warming so far, some 55 per cent.

It is important to appreciate the challenges facing those researching these changes. The 0.7 °C rise in average global temperature over the past century is well-established scientifically as are the increases in atmospheric concentrations of gases such as carbon dioxide (CO_2) and methane (CH_4), both of which are effective at trapping radiation within the atmosphere. Measuring and estimating sea level rise faces the particular challenge of the sheer scale of the oceans in three dimensions. Intense research efforts around the world continue to record and interrogate the vast quantities of data being collected by ever more sophisticated techniques. Academic debate refines arguments and leads to greater understanding of past and present processes as well as allowing more authoritative prediction.

Recorded sea level change since the late nineteenth century

The primary data source for sea level change has been tidal gauge records. There is an impressive set of data stretching back over most of the previous century. However, it is difficult to derive accurate global figures as the sampling network of gauges is spatially very uneven. Another complicating factor is vertical movements in the land where gauges are positioned.

Since 1993, nearly complete global measurements of sea level between latitudes 66° N and S have been made by highly accurate instruments carried on the TOPEX/Poseidon and Jason-1 and -2 satellite systems. Overall the data show a clear net rise in mean sea level and an accelerating trend.

Source of data	Time period	Observed sea level change
Tide gauges	1870–1935	0.71 +/- 0.40 mm yr^{-1}
Tide gauges	1936–2001	1.84 +/- 0.19 mm yr^{-1}
Satellites	1993–2003	3.1 +/- 0.7 mm yr^{-1}

▲ **Table 5.4** Rates of sea level change

Major research efforts are going into assessing the causes of the increasing rate of sea level rise. **The Intergovernmental Panel on Climate Change (IPCC)** has estimated the relative contributions of sea level rise from various possible sources.

Source	Estimated contribution (mm yr^{-1}) and uncertainty margins
Glaciers + ice caps	0.77 +/- 0.22
Greenland ice sheet	0.21 +/- 0.07
Antarctic ice sheet	0.21 +/- 0.35
Thermal expansion of seawater	1.60 +/- 0.70

▲ **Table 5.5** Estimated contributions to observed sea level change from various sources

> **KEY TERM**
>
> **IPCC** An international body comprising hundreds of scientists who are researching and regularly reporting on the nature, causes and consequences of climate change.

Estimating the effects of glacio-eustasy is very complicated but is improving as more reliable and accurate data becomes available. It is important to appreciate the need for uncertainty margins, positive and negative, when investigating aspects of the environment of the scale of polar ice sheets or global phenomenon such as sea level rise. These do not diminish the trend of rising sea level in any way.

Future trends in sea level change

Predictions of future glacio-eustatic change come with margins of error and those made in the past are often revised. This is standard scientific practice and does not mean that researchers' findings are invalid. The IPCC uses several scenarios when looking at what might happen to greenhouse gas emissions and these then have impacts on predictions of sea level change (Figure 5.7).

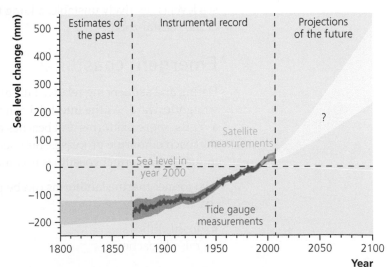

▲ **Figure 5.7** Sea level measurements and predictions

Depending on what happens to greenhouse gas emissions and other factors, such as the rate of removal of gases by natural means, the IPCC estimates that by 2100 global sea level will be around 0.2–0.6 metres higher than the 1980–99 levels. Changes experienced along different coasts will not all be the same. As we have seen, relative sea level changes depend on more than what is happening to sea level, such as tectonic adjustments and continuing isostatic adjustments. The sea levels that the people living at the end of the century will face depend on the ways that natural systems respond to anthropogenic and other **forcings**, as well as myriad human decisions. Feedback mechanisms in the Earth-atmosphere-ocean system are very complicated and modelling of trigger factors and what might happen to sea level when certain thresholds are crossed can only operate within a range of probabilities.

KEY TERM

Forcings Factors in any system that drive or 'force' a change. They can be of natural or human origin.

② Characteristic landscapes and landforms

▶ *What coastal features develop when there is a relative change in sea level?*

Sea level is an important control on coastal landforms and landscapes largely because where sea level is determines the flows of energy and materials through the coastal system. A change in sea level will result in alterations to all the coastal processes, sometimes simply modifying them but in other circumstances bringing about a complete change in the processes acting upon the shore. Feedback processes and the state of any particular part of the coastal system can represent significant modification in equilibrium. In geological and geomorphological time scales, change in sea level is relatively unstable, adding to the already highly dynamic nature of the coastal zone.

Emergent coastlines

Dating and sequencing relative falls and rises in sea level is not straightforward. As the intertidal zone falls lower, coastal landforms such as cliffs, shore platforms and beaches are left literally 'high and dry', beyond the reach of marine processes. Sub-aerial processes become more influential as former coastlines become 'fossil' coastal landscapes.

Two former marine landforms can be prominent along an emerging coastline:

- fossil cliffs
- raised beaches.

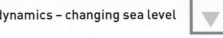

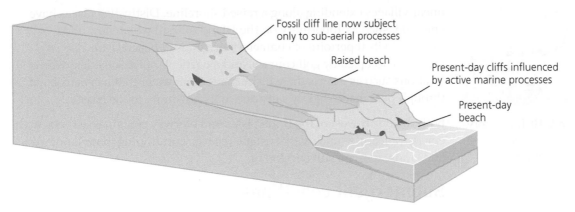

▲ **Figure 5.8** Landforms and landscape of an emergent coastline

Former marine cliffs mark the line of the coast when sea level was higher than it is today. They are particularly distinguishable when their geology resists sub-aerial processes, such as igneous and metamorphic rocks. Former sea caves and wave-cut notches can be recognised in some locations.

Raised beaches exist as relatively gently sloped areas made up of sandy soils, although some can be covered by **head**. As the last ice age of the Pleistocene drew to a close, periglacial processes were very active in the landscape. While it is the case that a relative fall in sea level does produce raised beaches, the word 'raised' may be misleading in some examples. Some fossil cliff lines and raised beaches were formed where they are found today, the result of marine processes operating during times of higher sea level.

In areas where steep relief dominates inland, the flat spaces that raised beaches represent, have been changed into important places by human activities. In Cornwall, farmers have traditionally exploited the sandy soils, which warm early in spring, for cultivating vegetables and flowers (especially daffodils). In the north-west of Scotland, both the mainland and the many islands of the Inner and Outer Hebrides have small,

> **KEY TERMS**
>
> **Raised beach** An area of former beach now stranded above present-day sea level.
>
> **Head** A deposit of clay and angular stony rubble formed by slow mass movements under periglacial conditions.

▲ **Figure 5.9** Emergent coastline in north-west Scotland

▲ **Figure 5.10** Tobermory, Island of Mull, Inner Hebrides, Scotland

linear villages extending along a raised shoreline. Distinctive places have emerged, such as Tobermory on the Island of Mull which now adds tourism to its repertoire of characteristics. Raised beaches are also locations occupied by golf courses, and a particular place characteristic of locations such as Islay is the whisky made from water that has percolated through **peat** made slightly salty due to its proximity to the coast.

Both the north-west of the British Isles and Norway contain coastlines with emergent features and, with isostatic recovery continuing, both landforms and landscapes are still developing.

Submergent coastlines

Around many coastlines, much coastal land has been drowned by the post-glacial rise in sea level. The majority of the world's coastlines are relatively young, only a few thousand years old. In some regions, the outline of the land has changed considerably during the Holocene.

Between the British Isles and the Netherlands, under the surface of what is now the southern North Sea, lies a drowned region. At the end of the Pleistocene, this region consisted of gently sloping hills, marshland and wooded valleys with a flourishing ecosystem. The area was inhabited by people migrating north as the ice sheets retreated. Rising sea levels progressively reduced the area of land, leading to one of the first examples of environmental refugees as people were forced out of the area. Today, this part of the North Sea is relatively shallow, in many places only some 20 metres deep and with many shallower locations and shifting sand and mud banks. Trawlers regularly bring up animal bones and remnants of stone tools.

Estuaries

Most estuaries around the world are drowned river valleys. In regions where there was little or no ice on the land, the relative rise in sea level has been significant because isostatic rebound has not occurred. Two types of drowned estuaries are recognised, rias and fjords.

Feature	Ria	Fjord
Cross-section shape	Flattened V-shaped	Narrow U-shaped
Depth	Can be quite deep at mouth (c.10 metres); shallows inland and in side creeks	300–400 metres common; can reach 1000 metres
Estuary sides	Steep	Steep
Altitude of estuary sides	<100 metres	Up to several hundred metres
Estuary plan	Dendritic	Relatively straight
Formed by	River	Glacier

▲ **Table 5.6** Rias and fjords compared

Rias

Rias are found in many regions of the world, such as the southern British Isles, Brittany, Galicia (Spain), the east coast of the USA and southern Chile. In these regions, towards the end of the Pleistocene and into the Holocene, large quantities of water flowed off the land as ice sheets melted. The rivers possessed considerable energy as their gradients reflected the fact that the sea level was well below that of today. The extensive river systems were typically dendritic as they collected water off the land.

> **🔑 KEY TERMS**
>
> **Rias** Drowned river valleys.
> **Dendritic** A network of stream and river channels similar to the branching pattern of a tree known as a dendritic pattern.

▲ **Figure 5.11** The Helford ria, Cornwall, looking towards the sea

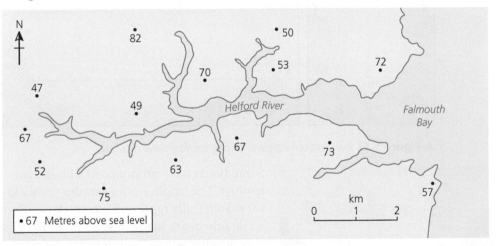

▲ **Figure 5.12** Plan view of the Helford ria, Cornwall

Where a river flows through a region of lower relief, a different type of drowned estuary forms. These are wide and shallow and tend not to have a dendritic plan, having relatively straight, parallel banks. In cross-section they are much broader and flatter. The estuaries of the east coast of England, such as the Thames, and the Potomac estuary in north-east USA are examples.

Fjords

In regions that were covered by ice, glaciers followed the lines of pre-existing river valleys. These 'rivers of ice' deepened, widened and straightened the former river-carved valleys, creating glacial troughs. As with rias, fjords developed as land-based ice melted and sea level rose. They are common along mid-latitude coasts such as Norway, eastern Greenland, western Canada and Chile.

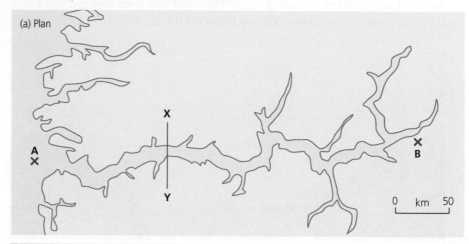

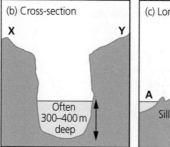

▲ **Figure 5.13** Plan, cross-section and long section view of a fjord

▲ **Figure 5.14** Entrance to Milford Sound fjord, New Zealand

Some fjords have pronounced sills at their mouths. The origin of this feature seems to vary. Some sills represent glacial deposition while others are made of solid rock. They may develop where the intensity of glacial erosion is reduced as the glacier fans out on to the former shore area or where it starts to float as it encounters the sea. Fjords can extend up to 150 kilometres inland and have spectacular waterfalls cascading down their sides.

Fiards are recognised as distinctive landforms in lowland areas that have been glaciated. These inlets are deeper than rias and do not have the plan nor the profile of fjords. They are characterised by having a large number of islands, such as the coastal region of south-east Sweden.

Rias, fjords and fiards are important to human activity, acting as sheltered anchorages. This role continues in some locations and these features can be magnets for tourists. The physical geography of these drowned coastal spaces has given rise to some distinctive places being created by human activities and representations. Many such locations offer sheltered anchorages for shipping and many fishing villages have grown up on narrow strips of flat land along fjords, for example in Norway and southern Chile. During the Second World War, Norwegian fjords such as Sogne were used as secure bases for the German navy, being readily defended due to the narrow entrance and width. Today, cruise vessels have replaced warships, with the fjords of Alaska, southern Chile, New Zealand's South Island and Norway receiving thousands of visitors each year. Part of Sognefjord has been deemed to be so special a place that it has been designated a World Heritage location (see Great Barrier Reef, page 106). As long as the depth of water allows large ocean-going vessels to enter, these drowned inlets can become important places as ports in the global system of shipping routes. The rias of Bantry Bay, Eire and Milford Haven, South Wales are destinations for oil tankers, as is the Thames Estuary which also acts as the import and export corridor for a diversity of goods. The process of globalisation relies on a network of international sea transport in which rias, fjords and fiards can become global hubs (see Chapter 6).

🔑 **KEY TERMS**

Fiard Drowned inlets of a former lowland glaciated area.

Dalmatian A coastline that is characterised by chains of islands running parallel to the mainland with deep bays and steep shorelines.

Drowned concordant coastlines

The term **dalmatian** is given to a particular coastal landscape affected by submergence. The term is derived from the Dalmatia region of Croatia along the Adriatic Sea, a region which was inundated during the Holocene. Concordant coastlines (page 52) can result in a series of parallel ridges and valleys lying along the coast. These often coincide with a geological structure of alternate anticlines and synclines, the ridges and valleys respectively. When sea level rises sufficiently to flood the coastal zone, the ridges of higher land are left as islands (Figure 5.15).

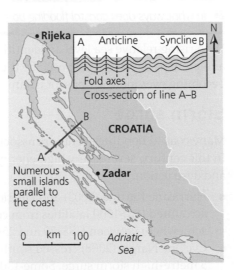

▲ **Figure 5.15** The Dalmatian coastal landscape, Croatia

🔑 **KEY TERMS**

Dendrochronology The scientific method of dating wood. It allows precise dating by counting the annual tree growth-rings. The pattern of annual tree rings differs each year, depending on the growing conditions at the time.

Pollen analysis The scientific analysis of microscopic pollen grains and spores preserved in sediments helps to reconstruct past environmental changes. Pollen grains are extremely resistant to decay, especially under anaerobic (without oxygen) conditions. Analysis of pollen from different layers of sediments allows past climate characteristics and any changes to be investigated.

Submerged coastlines

Around the world, there is a variety of coastal landforms and indeed entire landscapes lying below present-day sea level. For example off the coastline of the Bay of Biscay, western France, beach and sand dune systems exist at depths of between 100 and 200 metres. Around the coastline of south-west England, cliffs lie in 40–60 metres of water. Former shore platforms have been located around northern Australia at depths of 200 metres.

Submerged forests are present in the intertidal zone and just offshore along many coasts. The British Isles has a large number, mostly around the southern coastlines. They are found just a few metres below sea level, becoming exposed at particularly low tides. Evidence gathered from both **dendrochronology** and **pollen analysis** indicates that the trees flourished about 6000–5000 years ago. The climate had warmed after the last glacial, allowing a set of dynamic changes in vegetation as temperatures increased and a greater depth of soil formed. However, eventually, locations such as the mouth of the River Dovey in West Wales were flooded and the forest overcome by salt water.

These submerged features can be the result of sea level rise, tectonic subsidence or a combination of the two. The submerged forests in the British Isles are likely to have been drowned by both the glacio-eustasy of the Holocene and isostatic subsidence. As northern Britain rises because the weight of ice has been lifted, there is a degree of warping in the crustal rocks that results in subsidence in the south and east (Figure 5.16).

▲ **Figure 5.16** Rates of annual isostatic adjustment for Great Britain

3 Coastal flooding: risks and management

▶ *In what ways does coastal flooding pose risks and how can these be managed?*

With rising relative sea level an identifiable factor for the rest of this century and perhaps beyond, risks posed by the sea inundating coastal lands are becoming ever more serious for communities around the globe.

Storm surge

Many coastal flooding events are associated with storm surges (pages 7–8). In this century, several very high energy events caused considerable death and destruction:

- Hurricane Katrina (2005) and Hurricane Sandy (2012) together accounted for >1000 fatalities from coastal flooding in the USA and somewhere between US$179 and 250 billion in total costs.
- Cyclone Nargis (2008) crossed southern Myanmar, generating a 5 metre-high storm surge. Some 130 000 people died.
- Typhoon Haiyan (2013) struck the Philippines leaving about 800 people killed, missing or injured, most as a result of coastal flooding.

Northern Europe has a long history of severe coastal flooding, in particular as a result of storm surges. Medieval sources record repeated surge events (1099, 1421 and 1446) with associated high levels of mortality. Even with developing defence schemes including meteorological warnings, storm surges continue to bring destruction and in some cases death. A 1953 event led to the deaths of 1836 people in the Netherlands, 307 along the east coast of England and some 40 in Belgium. It was this event that was the catalyst for a change in attitude and approach towards the risks from coastal surges, leading to some dramatic engineering interventions such as the Dutch Delta Scheme and the Thames Barrier.

CONTEMPORARY CASE STUDY: NORTH SEA STORM SURGE 2013

In early December 2013, Storm Xaver developed off the coast of Greenland. As it crossed the North Atlantic its atmospheric pressure fell, causing high winds to circulate around the centre of the storm. During 4–5 December Xaver crossed northern Britain and by 6 December was tracking over southern Norway and Sweden. Along England's east coast, winds of about 60 km/hour were recorded and as the storm moved eastwards, the winds circulating anticlockwise around its centre blew from north to south, down across the North Sea (Figure 5.17).

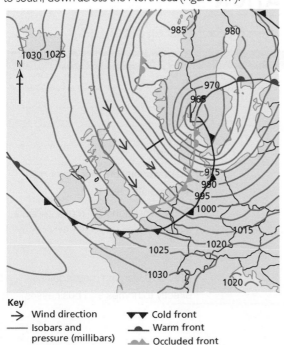

Key

→ Wind direction

—— Isobars and pressure (millibars)

▼▼ Cold front

●—● Warm front

▲▲ Occluded front

▲ **Figure 5.17** Atmospheric pressure at the surface for Storm Xaver, December 2013

Levels of high water rose well above the average for locations around the North Sea and even along the Channel.

Location	Time of high water	Observed high tide (metres)	Surge (metres)
Aberdeen	5/12/13 15:00	3.0	0.66
North Shields	5/12/13 16:15	4.04	1.25
Immingham	5/12/13 19:15	5.31	1.70

▲ **Table 5.7** The scale of tide and storm surge along Britain's east coast as Storm Xaver passed across the North Sea

Extreme high water levels were also experienced around the west coast such as at Liverpool and Ullapool.

Coastal flooding and its impacts

Some 2800 properties were reported flooded, along with an estimated 6.8 km² of agricultural land. About 18 000 people were evacuated in advance of the surge and 800 000 properties were protected by some 2800 km of flood defences. Insurance losses from flood damage were estimated at £100 million for the UK. There were numerous reports of coastal roads affected by seawater flooding and railway line damage was reported.

The Port of Immingham flooded, with severe damage caused to the dock infrastructure. A loss of around £115 million in damage and loss of business was reported.

Immingham is the UK's largest port in terms of tonnage handled, some 55 million tonnes annually. It is important strategically as it is the port of entry for considerable quantities of coal and biomass for power stations as well as oil, animal feed, grain, timber, road salt and containers. It is estimated that the surge event at Immingham represented a level of flooding likely to be experienced only once in 750 years.

With the considerable input of high wave and wind energy into the coastal systems, significant impacts were made to landforms along the coastal zone. Increased levels of erosion affected cliffs, especially those in relatively soft glacial deposits along parts of East Anglia's coastline. Sand dune systems experienced considerable erosion along their seaward lengths, with embryo, fore and some yellow dunes degraded. Gravel barriers and earth banks were breached and overtopped. Seawater then filled lagoons behind these and infilled with sediment wetland areas such as reed beds and grazing marsh.

▲ **Figure 5.18** Breaching of coastal defences caused by the December 2013 storm surge, near Salthouse, north Norfolk

ANALYSIS AND INTERPRETATION

Study Table 5.8, which contains data comparing two storm surge events for the east coast of England.

Impact	January 1953	December 2013
Major breaches of sea defences	120	4
Properties flooded	24 000	1400
Agricultural land flooded (ha)	65 000	6800
Deaths	307	0
People evacuated	32 000	18 000
Infrastructure put out of action for at least 24 hours	2 power stations 12 gas works 160 km roads 320 km rail track	No power stations nor major gas services affected Port of Immingham – major impacts
Flood warnings issued	0	64 severe >160 000 warning messages sent directly to homes and businesses

▲ **Table 5.8** Comparison of east coast storm surges

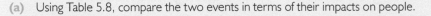
(a) Using Table 5.8, compare the two events in terms of their impacts on people.

GUIDANCE

Whenever there is the command to 'compare', it is important to compare in your response. This is best achieved by explicitly discussing the two events in the context of a particular impact rather than writing two separate accounts, one for each of the events. It is also helpful to distinguish between direct and indirect impacts. Mortality and evacuation would be examples of direct effects, with disruption to food supplies due to transport links being cut an example of indirect. The data clearly show a more significant set of impacts on people from the 1953 event than the 2013 storm surge. One very stark comparison is in the death toll, with just over 300 people killed in 1953 compared to no one in 2013. Contrasts in level of impacts are also seen in the number of properties flooded and number of people evacuated.

(b) Suggest why impacts on infrastructure were so different between the two events.

GUIDANCE

The data show that both power and transport facilities were adversely affected in 1953, whereas in 2013 only the Port of Immingham was impacted. One possible reason is that infrastructure has been made more resilient, that is more capable of withstanding extreme events such as flooding. Roads and railways may have been better protected in 2013 through drainage and perhaps by being placed on embankments above the flood level. The 1953 event saw a much larger area of the east coast inundated, caused by a great difference in the number of breaches in the sea defences. This would have caused many more stretches of road, for example, to be underwater. In 1953 there were likely to be more power stations and gas works, each relatively small in scale, and so more of the energy infrastructure was likely to be at risk. By 2013, energy production was concentrated into fewer but larger plants, either not located in a hazard zone or likely to be better defended. The significant contrast in the issuing of flood warnings meant that in 2013, time was available to put in place temporary flood defences.

One consequence of most hazard events is that a reappraisal is carried out as regards mitigation of the risks and adaptation to high energy events. So severe were the impacts of the 1953 surge that measures were taken to minimise future risks from storm surges.

(c) Analyse the strengths and weaknesses of comparing data across a time period of several decades.

GUIDANCE

The issue of comparing spatial patterns over time is common in geographical investigation. Inevitably it begs the question, 'Is like being compared with like?' in terms of the data. One example common in geography is the comparison of data referring to spatial units such as counties, towns or wards within a town. Care needs to be taken that such comparisons are dealing with units with the same boundaries. Data is collected in different ways over time. Which questions are asked in a survey such as the Census and how they are worded can affect the nature of the data. Ways of measuring something can vary, such as who counts as unemployed.

Despite these potential weaknesses, being able to compare across time is valuable. It can highlight trends which, if established, make it important to ask further questions about the reasons why the trend has been as it is. Changes in sea level, for example, are prompting questions about why this is taking place, what can be done about it and what impacts the changes are likely to have on our coastlines and on the communities living in the coastal zone.

Rising sea level – a creeping geohazard

Sea level rise is likely to be one of the most dramatic consequences of global warming as this century proceeds. Its hazard potential lies not in sudden dramatic events such as earthquakes or tsunami, but in its almost imperceptible 'creeping' presence of just a few millimetres a year which leads, cumulatively, to a very significant change in average conditions. The implications for low-lying coastlines, especially in locations where subsidence of the land is occurring, are potentially very severe.

The IPCC has identified the following assessments regarding coastal locations:

- sea level – this will rise by between 28 cm and 98 cm by 2100, with most likely rise being 55 cm by 2100.
- delta flooding – the area of the world's major deltas at risk from coastal flooding is likely to increase by 50 per cent.

Both these assessments are given with 'high confidence', which the IPCC scientists give when their interrogation of the data suggests that the evidence has a 'high degree of certainty'.

The risks arising from relative sea level rise are distributed globally, but in terms of the implications for people and the management of such risks, a clear and concerning divide exists based on the **vulnerability** of affected communities. This divide is essentially along economic lines, with vulnerability reducing with increasing access to resources such as money.

However, given the relatively long time scale over which this hazard will make itself increasingly felt, the assessment of risk differs considerably across both locations and people's resources. A farmer eking out a living on a low island in the Ganges-Brahmaputra delta at the north of the Bay of Bengal will have a very short-term perspective of necessity. Providing food on a daily basis, managing to send children to school and concern over the next cyclone season are much higher priorities than longer-term sea level rise. Many home owners on barrier islands along the east coast of the USA may well appreciate the issue of sea level rise but in terms of decision making are likely to be focused on defending their property from the next flood or hurricane season. Few local, regional or national politicians take a long-term view of issues such as sea level rise in terms of allocating resources, although this is changing. The effect of severe damage following high energy storms has highlighted the risks coastal areas face and measures are being taken where resources allow, mostly in wealthy locations (see Chapter 7).

Coastal settlements right across the hierarchy from village to mega-city are susceptible, especially those located on:

- estuaries and lagoon shores
- deltas
- drained marshland
- small islands lying close to sea level.

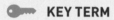

KEY TERM

Vulnerability Focuses on the ability of an individual, family or community to withstand exposure to, and risks from, out of the ordinary events, such as a geophysical hazard or serious illness.

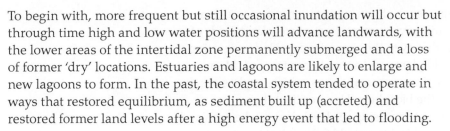

To begin with, more frequent but still occasional inundation will occur but through time high and low water positions will advance landwards, with the lower areas of the intertidal zone permanently submerged and a loss of former 'dry' locations. Estuaries and lagoons are likely to enlarge and new lagoons to form. In the past, the coastal system tended to operate in ways that restored equilibrium, as sediment built up (accreted) and restored former land levels after a high energy event that led to flooding.

A major concern to the management of sea level rise is that in many coastal locations, sediment supply into the coastal system has reduced. In many instances, this reduction is due to human activities. Dam-building many miles inland can severely disrupt sediment pathways. The reservoirs that develop upstream of a dam are low-energy environments, meaning that the river load brought into the reservoir settles out, including much of the suspended load. The Nile delta, home to about 40 million people living at an average density of 1000 per km^2, and where Egypt's agricultural productivity is highest, has experienced substantial declines in sediment inputs since the construction of the Aswan High Dam. The reductions both in water and sediment inputs across the delta has led to increases in severe erosion along the waterways crossing the delta and along the shoreline.

The reality is that sea level rise is unlikely to directly cause much mortality. As this century proceeds and average high tide heights increase due to sea level rise, risks for people living along the coast will increase, particularly when high tide and a storm combine. Devastating in different ways to death, is the loss of homes and businesses. Even in places where insurance is available, that is wealthier countries, many property owners are finding insurance impossible to buy in areas where coastal erosion and flooding have been experienced. As this physical characteristic of coastal communities changes, perceptions of such places are likely to alter. Once seen as attractive locations to live, increasing vulnerability to flooding and storm damage may lead to a down-grading of their appeal. Eventually, there may be an abandonment of sites and the inland migration of settlement.

Additional to flooding and erosion, sea level rise leads to salt water entering aquifers and thereby polluting fresh-water supplies. Domestic, industrial and agricultural activities rely on the availability of fresh water, which makes this unseen consequence of sea level rise serious. And it is not just human activities that are affected. Ecosystems are also at risk from the effects of rising sea level. Saline water will kill non-halophytic plants, deeper water for longer adversely affects salt marsh, mangrove and coral communities, and flooding and eventual submergence will eliminate ecosystems. The submerged forests caused by the post-glacial rise in sea level offer a view of what is likely to happen to many coastal ecosystems by

 KEY TERM

Aquifer Permeable rock that holds water, for example chalk.

the end of this century. The presence of non-natural ecosystems, such as intensive agriculture or the built environment, immediately landwards of the natural ecosystem, mean that even those species capable of migrating, such as non-flying insects, wild plants and animals, are trapped. Many coastal zones around the world host nature reserves, which emphasises the importance of sea level rise to the biosphere. How resilient these ecosystems will be to the threats from rising sea levels is yet to be seen and may become clear only when a critical threshold for survival is crossed.

Sea level rise and developed locations

Many urban centres in economically advanced countries are threatened by rising sea levels. Among the global elite cities, that is those that are leading in terms of activities such as business activity, information exchange, cultural experience and political engagement, several are threatened by rising sea levels: New York, London, Tokyo, Los Angeles and Amsterdam. Additionally, places such as Venice and Miami face severe challenges to the characteristics that help define them as distinctive places. Economically advanced countries have access to resources that allow them to identify, research, plan and invest in measures to deal with flood risks from sea level rise.

LOCATED EXAMPLE: FLOOD RISK AND ITS MANAGEMENT: LONDON

London's origins and development over many centuries owe much to its location along the banks of the Thames Estuary. The flood risk receded with the construction of structures such as the Embankment, which raised the level of the banks. Today, some 1.25 million people live and/or work on the floodplain. However, rising sea levels now mean that risks from seawater incursion, exacerbated by storm surges, require substantial management.

Specific risks from London's flooding include:

- Government – London is the focus of government activity both nationally and for the capital. Large numbers of key government offices are located on the floodplain either side of the estuary, such as the Houses of Parliament, the Whitehall district and the Greater London Authority, as well as boroughs such as Tower Hamlets and Newham. The Environment Agency in a 2007 report estimated that the loss in staff time in the civil service would cost some £10 million a day if flooding happened.

- Business – London is one of the small group of global leading centres. It is an international

financial centre linking New York with Tokyo in the 24-hour trading cycle. The Docklands area has attracted much service activity such as banks and insurance, and is in the floodplain. London's contribution to the UK economy is estimated at about £250 billion per year. London as a global tourist destination would be very disrupted by flooding. Eight power stations are at risk; shutting these would seriously disrupt power supplies.

- Infrastructure – over one billion journeys are made each year on the London Underground. Much of the system is in central London but perhaps the most significant impact would be on connectivity within the system. This would have severe impacts on journeys to work, school and college. Thirty-nine stations are at high risk from flooding and sixteen major hospitals are on the floodplain.

- Thames Gateway – this is Europe's largest regeneration project, with the target of 200 000 new homes by 2020. The area is mostly located in the floodplain.

Managing London's flood risks

The banks along the estuary have been successively raised, especially during the second half of the nineteenth century. Following the devastation due to the 1953 surge, research and planning eventually led to the building of the Thames Barrier, which became operational in 1982. This is a hard engineering solution consisting of massive gates that can be raised to provide a physical barrier to water coming up the Thames. When opened, the Thames Barrier had a life expectancy spanning up to 2030, and to February 2018 it has been closed 182 times. Not all of the closures are due to a flood risk from the sea as the Barrier also protects London from river flooding. The estimated life of the Barrier has been extended further into this century even with sea level rising.

▲ **Figure 5.19** Thames Barrier looking upstream with the financial district of Canary Wharf in the distance

Management of the flood risk is not restricted to the Barrier. The use of areas of farmland along the coasts of Kent and Essex to store floodwater from the sea, as well as temporary smaller-scale physical barriers, are additional measures.

Sea level rise and developing locations

Global warming and its impacts, such as rising sea level, are inextricably linked with the issue of the unequal distribution of global wealth and resources. Although all countries will suffer from the effects of global warming, it will be the poorest people who, as usual, suffer the most. Vulnerability is most acute for those who have the least. There is the likelihood that as this century proceeds, some locations in low-income developing countries will become uninhabitable and communities will be forced to leave as environmental refugees. Even within middle-income emerging economies, substantial segments of society may be extremely vulnerable to sea level rise. For instance, one UN report estimates that 40 million Indians will be at risk from rising sea levels by 2050. The important global hub city of Shanghai is threatened by rising sea level.

CONTEMPORARY CASE STUDY: THE MALDIVES

The Maldives is a group of some 1200 coral atoll islands located to the south-west of southern India.

Opportunities for human occupation

For centuries the islands have presented opportunities for human occupation, focused on farming the 10 per cent of the land area suitable for agriculture, fishing and exchanging goods with traders crossing the Indian Ocean.

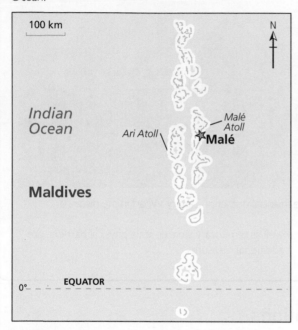

▲ **Figure 5.20** The main cluster of islands of the Maldives

In the later twentieth century, the Maldives engaged with the globalisation process via tourism, clothing and fishing for export (mostly skipjack tuna). The tropical climate and the potential of the warm coastal waters for activities such as exploring the underwater landscapes and ecosystems of the coral reefs have helped make the Maldives a popular long-haul destination for people from wealthy countries. Maldives tourism is aimed at the luxury end of the market based in resort developments.

Risks for human occupation

The physical geography of the Maldives offers both opportunity and risk. The total land area of 300 km² is surrounded by a 644 km coastline. The highest point is just 2.4 metres above sea level which means that the Maldives is one extensive coastal zone. Although the majority of the archipelago islands are uninhabited, sea level rise is a real and immediate risk. Increasing periods of inundation and enhanced wave erosion pose threats in various ways:

- Erosion of the coral reefs due to greater exposure to wave energy
- Seawater infiltrating the fresh water aquifers
- Land lost to sea reduces further the very limited area available for cultivation
- Loss of coral areas reduces fish breeding potential as reefs act as nurseries for eggs, larvae and young fish
- Coconut trees threatened by seawater contaminating the soil they grow in with salt.

Not only are the inhabitants directly threatened by rising sea levels, mainstays of their economy are as well. Tourism relies on the Maldives' informal representation as an island paradise. Loss of coral reefs (there have been some serious coral bleaching episodes) as well as iconic images such as palm and coconut tree-lined beaches, quickly translate to reduced numbers of visitors. A loss of income occurs both directly in hotels and restaurants but also indirectly among those in the support areas such as food supplies and airport staff. Unemployment rises and substantial flows of emigrants can form. Those most likely to move away are young adults, thereby depriving the islands of a vital element in the workforce but also splitting families and communities.

Management of risks

The resilience of the Maldives to risks from rising sea level is being tested as the country lacks economic and technological resources, but it has started to both mitigate and adapt. One approach has been 'land claim', such as the construction of Hulhumalé, a new artificial island based around an existing atoll, built from coral and sediment dredged from the seabed. On this land the 'City of Hope' is being built, protected by a 3 metre sea wall. The plans allow for a population of about 130 000 people by 2023 when the project is planned to be complete.

Resilience is helped by funds donated by foreign governments. Japan has paid US$60 million for a substantial sea wall surrounding the capital Malé. The wall is a hard engineering system of concrete tetrapods, concrete structures with four large spokes that interlock with each other. International agencies such as Mangroves for the Future are involved in projects on some atolls, such as developing nurseries for young mangrove trees which can then be planted to strengthen the level of defence against wave energy and storm surge. Cleaning lagoons of human rubbish and other debris aims to create as healthy an environment as possible so that the natural vegetation will be resilient to stresses caused by rising sea levels.

It might seem as if the risks posed to the Maldives would bring about a consensus and single-mindedness as regards approach. However, conflicts have arisen, in part because of the troubled political situation within the Maldives, but also due to the ongoing resort developments. There are plans to build tourist developments on more atolls and the threats to coral and marine ecosystems are seen by some as serious. The focus on main islands such as Malé and Hulhumalé as regards protection from rising sea level has resulted in less attention being given to more isolated atolls.

Evaluating the issue

▶ *Assessing the influence of sea level change on coastal landforms and landscape development.*

Possible contexts and criteria for the assessment

Relative sea level change can be either transgressive or regressive. Any discussion of the influence of changing sea level should include both and make it clear which type of change, rising or falling sea level, is being considered. For a comprehensive discussion, both types of change should be discussed, as by omitting one or the other the evaluation is incomplete.

Sea level change occurs over a number of different time scales. It is important to acknowledge the existence of long-, medium- and short-term change as the potential influence of each on coastal landforms and landscapes can be different. Considering the key concept of time is important when answering 'why' and 'how' landforms and landscapes have evolved.

As well as different time scales, another important context is the key concept of scale. The actual question highlights this to some extent by including both landform and landscape – two elements of the coastal zone at different scales. An individual landform might be a raised beach while a landscape could be a submerged coastline

such as the fjords of Norway or Chile. Spatial scale is a significant aspect of any analysis as by altering the scale it may be that different processes are highlighted and their relative influences assessed.

Assessing the influence of rising sea level

Present-day coastlines have experienced a largely stationary sea level only for about the past 6000 years. This is when the glacio-eustasy following the last major ice age came to an end. The implications of this situation are that because sea level change is global, all coastlines around the world are, to a greater or lesser extent, drowned. Sea level was some 120 metres below its current level when the last ice age was at its maximum, about 18 000 years ago. The Flandrian transgression (the term given to the post-ice age rise in sea level) has been responsible for a number of landforms and landscapes.

Because there has been an absolute increase in sea level since the last ice age at the global scale, a eustatic change, many coastlines have experienced transgressive conditions. Submergence is therefore a common experience

▶

and has, to a considerable extent, been responsible for the present-day shape of the continents. With the retreat of the continental-scale ice sheets, the outline of the Arctic Ocean, for example as defined by the coastlines of Alaska, northern Canada and Siberia, became what we see today.

At the regional scale, the formation of a Dalmatian-style coastline is down to the post-glacial sea level rise. Due to the concordant geological structure, ridges parallel to the coastline have been left as elongated islands while the valleys lying between the ridges have been flooded. Part of the Croatian coast has been formed in this way. However, this particular coastal landscape also owes its origins to tectonic forces which influenced the geological structure of anticlines (upfolds) and synclines (downfolds). In addition, pre-glacial weathering and erosion would have helped form the ridges and valleys.

At locations where glaciers, following the route of former river-cut valleys, met the sea, those valleys were made wider, deeper and straighter by the actions of the flowing ice. Once the ice melted and sea level rose, the lower sections of the glacial troughs were drowned to form fjords. Sea level rise has been a major influence on these coastal features but part of their evolution is due to the pre-glacial landscape. Rivers tended to cut along lines of relative weakness in the rocks and so present-day fjords reflect to some extent the landscape before the Pleistocene, the geological epoch when successive ice advances and retreats occurred.

Areas that were unglaciated have had river-carved valleys drowned by rising sea level. The rias of south-west England and Ireland, similar to fjords, reflect both pre-glacial processes as well as sea level rise.

At the scale of individual landforms, the coastlines of many mid-latitude regions possess shingle beaches. While coastal sediment can originate from several sources, research has indicated that it was the post-glacial rise in sea level that drove these beaches onshore, as areas previously exposed by the fall in sea level as ice built up on land were once again covered by the sea. These shingle beaches range in scale from relative small accumulations through to the vast feature of Chesil Beach (page 72).

Although permanently underwater today, a seabed landscape exists in the offshore zone along many coastlines. A common misunderstanding is that the seabed is a rather flat feature extending away from a beach or cliff. When relative sea level was considerably lower than it is today, the land extended much further beyond its current position. Across this area weathering and erosional process operated, rivers flowed and sediment was formed, moved and deposited. As sea level rose this landscape was lost from view but it continues to exist. Its importance is that flows of water and sediment in the coastal zone can be strongly influenced by the underwater relief. Canyons extending out into the Pacific from the Californian coast channel energetic flows of water and sediment that are key elements of localised and regional sediment cells.

Assessing the influence of falling sea level

It is known that sea level has risen and fallen over geological time, but with the passage of time at this very long-term time scale coastal landforms and landscapes either disappear altogether or are so modified by subsequent events and processes that they essentially disappear as recognisable features. However, the fall in relative sea level observed in some locations has formed some distinctive landforms and landscapes. The isostatic recovery of locations once under considerable thicknesses of ice during the Pleistocene has exceeded the glacio-eustatic rise. Emerging coastlines exist along coastlines such as in north-west Scotland, Scandinavia and the higher latitudes of North America. Cliffs, caves, notches, raised beaches and even stacks and stumps that were once under active marine influences have been lifted away from the effect of waves. These fossil features owe their origin to the change in sea level. However, now that they have emerged from coastal influences, sub-aerial erosion and weathering as well as mass movements are operating on them. As time passes, their marine features will gradually be modified. Cliffs are likely to reduce in steepness as material builds up at their base as it is no longer removed by wave action. Raised beaches will become subject to soil-forming processes and can be used for agriculture, albeit with sandy soils.

Some of the drowned features such as fjords also have raised beaches around their shorelines which indicates that some coastlines are the product of both rising and falling relative sea level. Such landscapes are useful reminders that what we see today is rarely the product of just contemporary processes nor of only one set of circumstances in the past. As with what are essentially human landscapes (settlements, for example), physical landscapes and indeed many landforms are the result of multiple influences.

Assessing the influence of other factors

Because of the scale of the recent eustatic rise following the last major glaciation and the isostatic recovery in regions such as northern Canada and Scandinavia, the influence of relative sea level change to coastal landforms and landscapes is significant. However, it is important to recognise that the coastal zone operates as a system within which several factors operate.

Given the extreme time scales involved, the role of tectonic events in the geological past can be hard to appreciate. They are, however, crucial in providing the geological framework within which individual landforms and landscapes develop. The chalk cliffs of southern England and northern France are the shape they are in large part due to the conditions under which the chalk formed, that is during the Cretaceous period, some 66 to 145 million years ago. The warm subtropical sea existing at the time across this region was home to the marine organisms (phytoplankton) whose skeletons, made of calcium carbonate, make up the majority of chalk rock. The porous nature of chalk allows water to percolate through and so not be as worn away by flowing water over its surface as happens to impermeable clay for example. The collision of the African and Eurasian Plates released vast amounts of energy during the Alpine orogeny, some 2 to 65 million years ago. Even over a thousand kilometres away, rock strata were folded and fractured, the influence of which can be seen in the coastal landforms either side of the Channel.

The spectacular fjord landscape of Norway's and Chile's coastlines is not only the result of submergence but also the shear strength of the geology making up the towering steep rock faces. That some igneous and metamorphic rocks are capable of supporting almost vertical free faces is a key factor in the development of the fjord. The rias of south-west Britain are also due in part to the geology from which they are formed. The weathering and erosional process that carved the dendritic river networks interacted with this geology to create a more subdued landscape. Even if ice had covered these locations, deep glacial troughs would not have been formed, rather a gently undulating landscape as seen in large areas of northern Europe and America.

Increasingly, the influence of human activities affects coastal landforms and landscapes. Although the post-glacial rise in sea level has been a key factor in the development of shingle beaches such as along the coastline of Cardigan Bay, Wales, human interventions in the form of sediment removal and the construction of groynes can be locally significant.

Arriving at an evidenced conclusion

What is the extent of the influence of sea level change on coastal landforms and landscapes? Because of the events of the Quaternary, the most recent geological period, sea level change, both rising and falling, has had very significant influence. It is hard to avoid these influences around many coastlines, especially in the mid-latitudes such as north-west Europe. The influences of submergence and emergence are present at a variety of scales, from landscapes through to individual landforms.

However, events and processes across geological time have a role to play, such as in the formation of the very rocks which make up the coastline. Additionally, present-day processes are modifying features. Wave action is active on slopes once at some distance from the shore; estuaries such as rias receive sediment from the rivers that flow into them. In some locations, human activities play a part such as with sediment removal or coastal protection. Coastal landforms and landscapes are the product of the interaction of a range of factors across a wide range of spatial and temporal scales.

 KEY TERM

Orogeny Both a period of time and the mechanisms which form mountain chains such as the Himalayas and the Andes.

Chapter summary

✔ The idea of relative sea level is important as this describes the rise or fall of mean sea level compared to the land. Relative changes are either transgressive – drowning of the land – or regressive – emerging of more land.

✔ Two types of relative change are recognised: eustatic changes in sea level are global, while isostatic changes in the land level tend to be regional or local. Both eustatic and isostatic can arise from natural forcings such as sea level fall during an ice age or land rising due to tectonics. Both eustatic and isostatic can arise from human factors such as sea level rise due to global warming or land subsidence due to ground-water abstraction. Eustatic and isostatic changes associated with the ending of the last glacial period have been very significant in the development of coastal landforms and landscapes. A relative fall in sea level produces features such as fossil cliff lines and raised beaches, while a relative rise in sea level produces features such as estuaries (rias and fjords) and a Dalmatian coast line.

✔ Global warming is having a significant impact on sea level change as regards both melting of land-based ice and the thermal expansion of water. The threat of coastal flooding is a significant and increasing risk in many parts of the world due to storm surge and the rise in relative sea level. Relative sea level rise is likely to affect millions of people as this century proceeds, with the poor least able to cope due to their high levels of vulnerability.

Refresher questions

1 Define the terms 'eustatic' and 'isostatic'.

2 Explain how change in the hydrosphere can impact on sea level.

3 Outline how human activities can cause subsidence at the coast.

4 Explain how global warming leads to a relative rise in sea level.

5 Explain the formation of landforms caused by regressive conditions.

6 Explain the role transgressive conditions play in the formation of beaches and 'slope-over-wall' cliffs.

7 Describe and explain the circumstances that favour the development of a storm surge.

8 Outline ways in which impacts from rising sea level can be managed.

Discussion activities

1 In pairs, discuss the significance of the 'fossil' nature of much of the shingle sediment in the coastal zone. What are the implications for the ways this sediment might be valued and managed?

2 Study Figure 5.21.

Describe and suggest reasons for the relative change in sea level between Times 1 and 2. Draw a similar diagram to help you describe and suggest reasons for the relative change in sea level when eustatic change is greater than isostatic change. Discuss the possible effects of this situation on a cliffed coastline, a sand dune system and an estuary with extensive salt marshes.

3 Research the current predictions for sea level rise for the rest of this century. The IPCC website (www.ipcc.ch) publishes reports on the latest scientific estimates. Discuss the possible implications of the various levels of sea level rise on aspects of the coastal zone such as sediment movement, the equilibrium of sand and shingle landforms and the equilibrium of cliff profiles. Use the detail of an Ordnance Survey 1:25 000 scale map to set your discussion in a real-world context. Sketch how the coastline might look depending on the lower, middle and upper estimates of sea level rise.

4 Using atlas maps and Google Earth, identify three coastlines from countries at different points along the development continuum that are particularly at risk from contemporary rising sea level. Compare your list with those of others and suggest appropriate measures that could be taken to mitigate and adapt to higher relative sea levels.

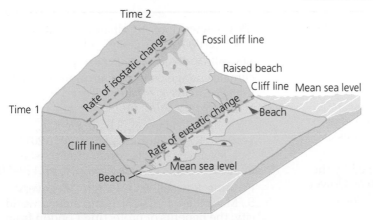

▲ **Figure 5.21** The relationship between isostatic and eustatic change

FIELDWORK FOCUS

A *Examining a shingle beach to assess how active present-day processes are*. If longshore drift is active, then one might expect a difference in both size and shape of sediment in the direction of drift. The degree of difference in sediment could indicate active present-day processes or suggest that the sediment was essentially a relict feature.

B *The development of landforms such as fossil cliffs and raised beaches*. A potentially useful starting point is the 1:25 000 Ordnance Survey map series, with its contour interval of only 5 metres. Drawing cross-sections of transect lines running inland from the high water mark at locations along a stretch of coast may reveal distinct breaks in slope as can be seen in Figure 5.8. These can be surveyed in the field where access is possible.

C *Investigating the influence of past events, such as isostatic change, by measuring pH levels and soil texture along a transect*. Sandy soils are indicative of raised beaches and where there is a sequence of sites at different heights, as in Figure 5.8, comparisons can be made.

In the illustration, sampling could be undertaken in the area below the recently abandoned cliff line, on the raised beach and on the flat land just below the fossil cliff line.

D *Interviewing samples of people in order to investigate perceptions of risk from a rising sea level. Questionnaire surveys could be used to assess levels of knowledge of local risks and how and why different groups possess varying perceptions of risks*. This could be tied in with a study of place characteristics and how these might be changing. Collecting data about the background of the respondents is valuable, such as how long they had lived there. Interviews could be conducted with local councillors, planners and agencies and the Environment Agency.

E *A detailed local-scale survey of a small settlement threatened by flooding*. A base map showing individual streets and buildings could be used to plot various adaptations individuals and the community have taken, such as raised thresholds to front doors, fittings for temporary flood barriers across doorways and larger-scale flood gates across access points from the sea.

Further reading

Bird, E. (1993) *Submerging Coasts*. Chichester: Wiley

Church, J.A., Woodworth, P.L., Aarup, T., Wilson, W.S. (eds) (2010) *Understanding Sea-Level Rise and Variability*. Chichester: Wiley-Blackwell

Classic Landform Guides published by the Geographical Association such as Gower Coast, North Devon Coast and West Dorset Coast

Environment Agency UK – environment-agency.gov.uk

IPCC – www.ipcc.ch/pdf/assessment-report/ar5/wg2/WGIIAR5-Chap5_FINAL.pdf

Masselink, G., Russell, P. (2013) 'Impacts of climate change on coastal erosion', *Marine Climate Change Information Partnership, Science Review 2013*, pp.71–86

Nicholls, R.J., Hanson, S.E., Lowe, J.A., Warrick, R.A., Xianfu, L., Long, A.J. (2014) 'Sea-level scenarios for evaluating coastal impacts', *WIREs Climate Change*, 5, pp.129–50

Penning-Rowsell, E.C., Haigh, N., Lavery, S., McFadden, L. (2013) 'A threatened world city: the benefits of protecting London from the sea', *Natural Hazards*, 66, pp.1383–1404

CHAPTER 6

The significance of coasts for human activities

Lying between the sea and the land, the coastal zone attracts some human activities but restricts and even repels others. Tidal movements of water mean that intertidal locations tend not to suit permanent land uses such as settlement, industry and agriculture. However, landward of the high tide level, land uses are attracted to the coast. The resources contained within the coastal zone can be exploited for food, energy, tourism and recreation and for land to build on. The exchange of goods and people across oceans has led to some coastal spaces developing significant meanings as economic, social and political places. This chapter:

- explores the growth in population and settlements in coastal zones
- investigates resources in the coastal zone
- analyses the importance of the coastal zone in the globalisation process
- assesses the value of the coastal zone for human activities.

KEY CONCEPTS

Globalisation The multiple interconnections and linkages between nations, groups of people, businesses and individuals which make up the modern world system. The key element is the integration of human activities across the globe. In this, the coast is no longer the barrier it once was as goods and people criss-cross the world's oceans through ports. The growth in globalised economic activities is a major factor in the development of so many large settlements in coastal locations.

Interdependence This exists when and where human societies have relations among each other in economic, social and political areas. Interdependence is now worldwide and part of globalisation. The coastal zone plays an important role in this, for example in the exploitation of coastal resources such as oil and gas. Most human societies have become very reliant on the fossil fuel economy, where accessible oil and gas resources are exploited and then traded globally.

Inequality The unequal distribution of resources and opportunities. The attributes of the coastal zone vary spatially so that some locations have a rich resource base, such as fish or sandy beaches for tourism, while others are lacking in resources. Some coastlines offer opportunities for port development while others lack safe anchorages. Both globalisation and interdependence raise concerns about 'winners' and 'losers' in terms of economic, social and political power. The coastal zone is where advantages and disadvantages can be seen.

Sustainability At its most basic this refers to the idea of the use of something that can continue in ways that respect the needs of future generations. A key aim is to ensure the integrity, productivity and health of environmental and economic systems. The intention is to achieve a balance between supply and demand without having adverse impacts on the physical and economic environment. The use of coastal resources such as fish stocks for food or coral reefs for tourism pose questions about sustainability which their management attempts to address.

The growth of population and settlements

▶ *How have human activities developed along coast lines?*

Coastal areas have long been important to the distribution of population. Some of the earliest archaeological evidence for human habitation has been discovered at coastal locations. The potential for gathering food such as shellfish, fish and wild fowl as well as plants was exploited by early societies around the globe. Over the centuries the association between proximity to the sea and density of population has become stronger. Coastal populations are growing faster than the national average in virtually all countries with a coastline.

About half of the world's 7.5 billion people live within 200 kilometres of a coast, an area representing some 10 per cent of the Earth's land surface. Of the inhabited continents, only in Africa do more people live in the interior than in the coastal zone. Among Asian countries, some 1.6 billion people live within 100 kilometres of the sea. Over half the Chinese population inhabit the coastal provinces, with densities along much of the coast averaging 600 persons per square kilometre. By 2025 some 75 per cent of the residents of the USA are expected to live in coastal areas, which together account for about 17 per cent of the land area (excluding Alaska and Hawaii). An estimated 200 million Europeans (out of about 680 million) live within 50 kilometres of the coast.

There are, however, significant latitudinal variations in the attractiveness of coastlines to human settlement. High latitude coastlines such as northern Russia and southern Chile are sparsely populated. Generally, coastlines in the mid- and low latitudes are more densely populated. Fertile, low altitude coastal plains, deltas and estuaries are the most densely populated coasts. The Nile and the Chang Jiang (Yangtze) deltas and the Rio de la Plata (River Plate) estuary have attracted millions of people.

The importance of scale when describing coastal populations

At the global scale, continental interiors are sparsely populated when compared to coastal locations. Looking at a scale such as that of the entire southern coast of England, a picture can emerge of relatively high population density along the coast. However, by zooming in to investigate at a smaller scale the picture can alter. Locally some stretches of coast have relatively few people living along them. A rocky coastline with steep cliffs and limited access to the sea offers few opportunities for settlement and economic activity. For example the Dorset coastline between Weymouth in the west and Swanage in the east has a comparatively low population density compared with the south coast of England as a whole.

People living close to sea level

The attraction of the coast for human activity is readily identified in the statistics showing numbers living close to the coast. However, perhaps an even more significant perspective is to consider human activity in relation to low-lying coastal locations. As the impacts of climate change increase in intensity, rising sea levels pose major uncertainty for human activities close to sea level.

Some 5 per cent of the world's population, about 375 million people, live within 5 metres of sea level. There is, however, great variation among countries.

Country	% population living <5 metres above sea level	Total population (millions)	Number of people living <5 metres above sea level (millions)
Australia	5	24.1	1.2
Bahrain	34	1.4	0.5
Bangladesh	9	163.0	14.7
Belgium	11	11.4	1.3
Benin	12	10.9	1.3
China	7	1400.0	98.0
Côte d'Ivoire	4	23.7	0.9
Denmark	16	5.7	0.9
Egypt	22	95.7	21.1
Indonesia	7	261.1	18.3
Italy	6	60.6	3.6
Japan	13	127.0	16.5
Netherlands	59	17.0	10.0
Senegal	10	15.4	1.5
Tunisia	8	11.4	0.9
UK	5	65.6	3.3
Vietnam	37	92.7	34.3
World	5	7500.0	375.0

▲ **Table 6.1** Percentage of population living <5 metres above sea level for selected countries, 2017

The Netherlands stands out as well over half its population live close to or below sea level. It is important to appreciate the contrast between looking at the percentage of a country's population and the absolute number of people. Bahrain and Vietnam have similar proportions of their inhabitants living close to sea level, 34 and 37 per cent respectively. The actual number of people these percentages represent is starkly different, half a million in Bahrain but just over 34 million Vietnamese.

KEY TERMS

Conurbation A large urban area in population and areal terms, developed by the merging together of previously separate towns and cities.

Infrastructure The term given to fixed capital investment in transport networks, utility grids for example water and power, housing, hospitals, schools and so on that are essential to the effective functioning of an economy and society.

As with any issue facing countries, people's vulnerability depends on the resources they have available to respond to and deal with impacts arising from the issue. Rising sea levels are likely to have their greatest impacts on poorer countries as mitigation and resilience require economic and technological capabilities. Just as inequalities occur between countries, it is the same within a country, as it is the better-off who generally are able to cope with change.

Urban coastal development

The attraction of coastal locations globally is clearly seen in the locations of the major metropolitan regions or **conurbations**.

Most of the urban centres occupy sites on, or very close to, the coast. Fourteen of the USA's twenty largest conurbations are coastal (Figure 6.1). Of China's 456 officially designated municipal cities, 305 are in the coastal zone. In Latin America and the Caribbean, 57 out of 77 major cities have coastal sites.

Infrastructure developments go hand in hand with the growth of coastal populations. Many coastal cities have evolved in association with port activities. As the import and export of goods grows, various feedback loops operate that encourage multiplier effects and, therefore, more growth (Figure 6.2).

▲ **Figure 6.1** Waterfront development, Boston, Massachusetts

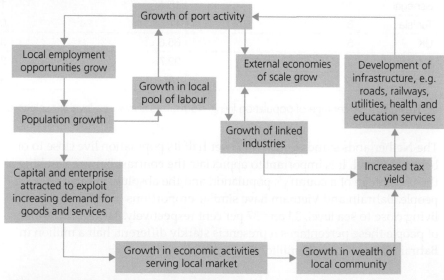

▲ **Figure 6.2** The growth of port activity

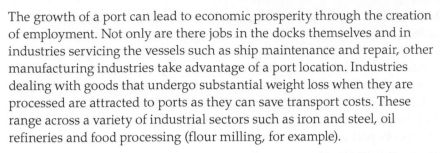

The growth of a port can lead to economic prosperity through the creation of employment. Not only are there jobs in the docks themselves and in industries servicing the vessels such as ship maintenance and repair, other manufacturing industries take advantage of a port location. Industries dealing with goods that undergo substantial weight loss when they are processed are attracted to ports as they can save transport costs. These range across a variety of industrial sectors such as iron and steel, oil refineries and food processing (flour milling, for example).

With population growth, demand increases for goods and services for these people, including food which stimulates agriculture in the region. To support the development of residential, industrial and service sectors, infrastructure is improved. All the while local and national tax revenues grow, allowing further investment in areas such as schools and medical facilities, roads and sewerage systems.

However, the physical developments associated with a port can also increase the pressure on coastal environments and ecosystems.

River ports, for example Liverpool and Rotterdam, developed inland from the open sea along tidal stretches of estuaries. Sea ports grew along stretches of open coast, for example Los Angeles. As soon as rudimentary wharfs and bridges are constructed, the estuary system is altered. By the seventeenth century much of the tidal stretch of the lower River Thames was already lined with an artificial shoreline of wooden walls. The original London Bridge with its large number of stone and brick piers reduced the flow of water and so affected estuary processes such as the transport and deposition of sediment.

By the mid-eighteenth century ports were constructing docks landward of lock gates. This avoided the effects of tides on moored vessels and allowed walls to be built to keep cargoes secure in warehouses. Large areas of floodplain were taken up with dock premises and the associated industry, railway and road links and housing. As the Port of London expanded eastwards, areas either side of the Thames such as marshland on the Isle of Dogs were built over. Not only was the hydrological cycle affected but also fluvial and coastal processes were altered. The flow of water was constrained between artificial banks and navigation channels maintained by dredging sediment. Meanwhile jetties and breakwaters modified longshore drift of sediment.

Changing coasts and changing places

Sometimes, physical processes are a significant contributory factor in changing a place's characteristics and status. In north-west England, Chester was the premier port up until the early eighteenth century. However, as the River Dee estuary became increasingly silted up due to its low wave energy environment and high sediment load carried by the river, ships were no longer able to navigate safely up the estuary to the city's wharfs.

Meanwhile, a small fishing village on the east bank of the River Mersey called Liverpool, just some 25 kilometres away, began to develop as a port. One major advantage for Liverpool was the way the Mersey estuary shape, greater water depth and fast-flowing currents kept this stretch of the coast relatively free from sediment. As the size of ships steadily grew, so the shift in fortunes of Chester and Liverpool accelerated. By the 1970s, Liverpool itself began to lose shipping trade as vessel size increased yet further. Today Liverpool's port activities are focused at the actual river mouth while docks upstream have closed.

 # Resources in the coastal zone

▶ *What resources can the coastal zone offer to human activities?*

The coastal zone possesses a wide range of resources that humans have come to value. The use made of coastal resources has changed through time and continues to be a dynamic factor in human interaction with the coast.

What is a resource?

There is no single agreed definition of a resource but in general it is something that satisfies human wants and needs. Something is not a resource until human society defines it as such. Some resources become very significant but then decline in importance as something else takes their place.

A basic distinction can be drawn between natural and human resources. The former includes substances, organisms and properties of the physical environment such as wind and space, together termed natural capital (page 88). These are valued because people perceive them to be useful in satisfying wants and needs. In coastal zones the value of natural capital is increasingly recognised through the use of ecosystem services (page 88). The value of a sand dune or salt marsh system to aid flood prevention is increasingly recognised; mangroves and coral reefs are seen as important breeding and nursery locations for fish. Human resources, human capital, are features of the human population such as numbers of people, their abilities and skills.

A valuable way of thinking about resources is to assess how renewable or non-renewable a particular resource is. Renewable resources are also known as flow resources, non-renewable as stock resources.

A characteristic of stock resources is that they tend to have limited spatial availability. For example, minerals such as copper and gold are only found in specific geographical locations. Others are spatially dispersed, such as sand and gravel. However, some resources cannot be easily placed in a particular category. Oil and gas, for example, tend to be located in particular

KEY TERMS

Renewable resource This is capable of regeneration within human time scales.

Non-renewable resource This cannot be replenished on a time scale relevant to humans.

regions such as the marine areas of the Gulf of Mexico and the North Sea. Within such regions, individual oil/gas fields can extend over considerable distances. This is another example of the scale of analysis being an important factor to consider.

The sustainability of resources

Allocating a resource to a category such as renewable is rather simplistic. With the technological advances of recent decades, and anticipating more to come in the next few, it is perhaps more appropriate to think in terms of a 'resource continuum' (Figure 6.3).

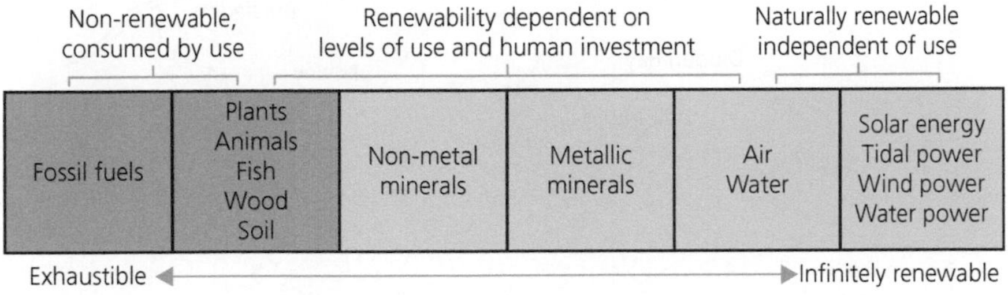

Non-renewable, consumed by use		Renewability dependent on levels of use and human investment		Naturally renewable independent of use	
Fossil fuels	Plants Animals Fish Wood Soil	Non-metal minerals	Metallic minerals	Air Water	Solar energy Tidal power Wind power Water power

Exhaustible ◄─────────────────────────────► Infinitely renewable

▲ **Figure 6.3** The continuum of resource sustainability

It is not just technology that influences resource use and degree of renewability. Human attitudes and perceptions play very significant roles. Increasing concerns regarding the stresses being placed on resources have resulted in a focus on management by individuals, groups and societies. Essentially such management tries to ensure that use of a flow resource allows it to regenerate itself, such as fish stocks. In the case of stock resources, the aim is to prevent supplies running out completely through improved recovery methods such as advances in mining and drilling technologies and recycling.

Coastal zone resources: land from the sea

The demand for land space and access to the sea has led to significant human intervention in natural coastal systems. Estuaries and coastal lowlands have come under particular pressures, especially in the past two hundred years. In some coastal locations little suitable undeveloped land remains, with the consequence that intertidal areas are identified as offering much potential for construction. Tidal flats, sand dunes and salt marsh can be engineered to provide dry land. Embankment construction stops the sea from entering, allowing drainage and consolidation of exposed sediments so that building can proceed. Both Southampton and Singapore have witnessed extensive conversion of intertidal areas into dry land (Figures 6.4 and 6.5).

Singapore has undergone substantial population growth over the past 50 years. Coupled with its considerable economic growth, demand for building land has been persistent (Table 6.2).

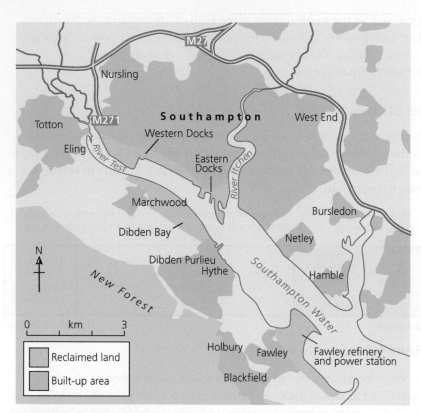

▲ **Figure 6.4** Intertidal reclamation in and around Southampton

Year	1970	1980	1990	2000	2010	2017
Total population (000s)	2074	2414	3047	4028	5077	5612
Population density (people/km²)	3538	3907	4814	5900	7146	7796

▲ **Table 6.2** Total population and population density Singapore, 1970–2017

Large-scale reclamation projects have added some 50 km² of land to Singapore's area and the process is continuing. Some of the reclamation has made land out of water 20 metres deep (Figure 6.5).

Land reclamation in the Netherlands

Some 10 000 years ago, much of the country now known as the Netherlands was an area of marshland, tidal inlets, tidal flats and sand dunes. In some locations peat accumulated. Early settlers built raised mounds called 'terps' where peat had accumulated above the flood level. Under Roman influence, causeways, canals and harbours were built. The first dykes (embankments) were recorded in the tenth century. The following centuries were times of give and take between land reclamation and water inundation, when, on balance, more land was lost back to the sea that was reclaimed.

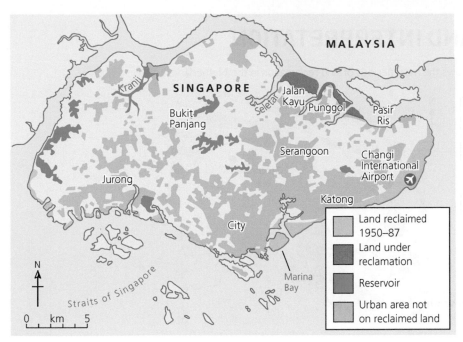

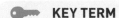

▲ **Figure 6.5** Coastal reclamation in Singapore

The twentieth century saw a large-scale application of technology to land reclamation. Schemes such as the partial draining of the Zuider Zee created the enclosed Ijsselmeer and its dry **polders** (Figure 6.6).

The Delta Project in south-west Netherlands witnessed immense efforts to bring about safety from coastal flooding and a greater security to many locations lying below sea level (Figure 6.7).

🔑 **KEY TERM**

Polders Flat areas of land reclaimed from the sea, often lying below sea level.

▲ **Figure 6.6** Polder landscape near Lelystad, Netherlands

▲ **Figure 6.7** Dam gates in the Delta Project at the mouth of the River Rhine

ANALYSIS AND INTERPRETATION

Study Figure 6.8, which shows the main periods of land reclamation in the Netherlands.

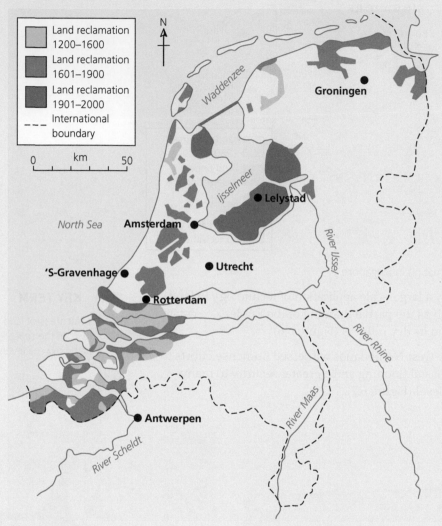

Figure 6.8 Land reclamation in the Netherlands, 1200–2000

(a) With reference to Figure 6.8, describe the pattern of land reclamation in the years from 1200 to 1600.

GUIDANCE

When describing any pattern it is helpful first to consider the overall distribution. The map indicates that between 1200 and 1600, land reclamation was focused in the south-west of the Netherlands. In particular, reclamation was undertaken at the mouths of the Rivers Maas, Rhine, Scheldt and Ijssel. There were also a couple of areas of reclamation in the north. It is important to extract actual detail such as names and locations from a resource when the question explicitly states 'With reference to …'

(b) With reference to Figure 6.8, compare the pattern of land reclamation for the years 1601 to 1900 with that for the years 1901 to 2000.

GUIDANCE

The command word 'compare' requires a response to include explicit comparison. The essential comparison is between more widespread and smaller areas of reclamation in the first period, and fewer but individually larger projects in the second time period. Between 1601 and 1900, land reclamation was carried out along a linear band parallel to the North Sea coast, stretching as far as some 20 kilometres inland. During the nineteenth century, three large areas of reclamation were undertaken in the Ijsselmeer in the north of the Netherlands with a fourth area along the River Ijssel between Rotterdam and the North Sea.

(c) Explain why the implementation of coastal land reclamation is often a contested activity.

GUIDANCE

This is a relatively open-ended question inviting consideration of the different perspectives that the various players (stakeholders) involved in a reclamation project might have. It can be useful to employ a spider diagram to help structure a response and to highlight the diversity of players affected by coastal reclamation. Reclamation is often driven by economic factors and population pressure. Gaining space for land uses such as manufacturing industry, port facilities, airports, offices and residential has been important in places such as Southampton, Singapore and Hong Kong. There may be a contest for space among these uses but more likely are the opposing views of those in favour of the reclamation compared with those against the change. The latter diverse group of players are likely to gather around the belief in the conservation of the natural environment. Questions raising concern about the loss of the 'sense of place' of undeveloped spaces are important. The overall need is to emphasise the competition for land and to recognise that different players hold contrasting attitudes.

Energy resources – hydrocarbons

The extension of sovereignty up to 200 nautical miles offshore has stimulated the exploitation of hydrocarbon resources, oil and gas. Countries offered, for lease or sale, exploration rights in blocks in the zone extending to the edge of the continental shelf. Today locations such as the Gulf of Mexico, Persian Gulf, Bight of Biafra in West Africa, and the North Sea, are major contributors towards oil and gas production. Such is the demand for hydrocarbons that few areas that have the geological potential to hold reserves are left unexplored. And as technology advances, drilling in deeper and rougher water becomes practical.

Both off- and onshore facilities develop to service the energy industry. Terminals where hydrocarbons are transferred, storage tanks, pipelines and refineries are built. The use of supertankers (Very Large Crude Carriers or VLCCs) has particular physical demands such as a water depth of thirty metres or more and not less than two kilometres of open water to turn in.

Inevitably, hydrocarbon exploitation has both positive and negative impacts. Advantages include wealth creation both through the employment generated, the value added to the raw materials such as the manufacture of chemicals from oil and the tax revenues generated. The latter allow local, regional and national governments to invest in infrastructure, enabling a higher standard of living for people. However, disadvantages arise when oil spills into the environment at levels that overwhelm ecosystems and disrupt food chains and webs. Docks and refineries tend to be large-scale structures that visually intrude into the landscape and can cause significant light pollution at night. There can be socio-economic disruption to local communities when the often substantial amounts of money brought into a place by hydrocarbon companies greatly distorts the local economy. During downturns in the hydrocarbon sector, for example when oil prices drop to low levels, there can be significant hardship in local communities as people are laid off and spending and investment dry up, as has occurred in north-east Scotland due to fluctuations in the North Sea oil and gas industries.

Oil spills

The release of oil into the environment comes about in various ways. Dramatic spills result from tanker wrecks or accidents involving drilling or production rigs, but these account for only five per cent of leaked oil. Most comes from spills during routine operations such as when oil is transferred between tanker and terminal. However, extreme events result in very serious impacts when they do occur.

CONTEMPORARY CASE STUDY: *DEEPWATER HORIZON*

In April 2010, the oil rig *Deepwater Horizon,* located 40 miles off the Louisiana coast in 1500 metres water depth, exploded. Eleven workers were killed and a further seventeen injured. The 'blow-out' device, designed to prevent high-pressure oil and gas from blasting up the drill pipe, failed (Figure 6.9).

It took 87 days before the wellhead was finally sealed preventing oil from gushing out into the Gulf of Mexico.

The amount of oil released is disputed; there is a ten per cent range +/- uncertainty around the estimate of 4.9 million barrels (780 000 m³) from the US Government. However assessed, the disaster was the largest oil spill in history. At its maximum, some 180 000 km² of the Gulf was affected and just over 1600 km of shoreline was polluted.

▲ **Figure 6.9** *Deepwater Horizon* rig after the 2010 explosion

What happens to oil when it spills into salt water is complex. It depends partly on the type of oil and on the sea and atmospheric conditions. Generally slick development goes through three phases:

1 Initial dispersion under influence of gravity (0–5 minutes)
2 Viscous **advection** (up to 40 hours)
3 Surface tension spreading (up to 150 days).

Processes of evaporation, solution, emulsification, oxidation and aeration take place and eventually, when the oil reduces to low concentrations, microbes biodegrade it.

Direct, short-term effects can be both visually and ecologically dramatic. Any impact on one **trophic level** will have significant effects throughout an ecosystem. Mass mortalities of some species result in a loss of biodiversity. Levels of resistance to oil vary among species depending on characteristics such as feeding behaviour. Diving birds such as cormorants and guillemots can be severely affected.

Oil spills on coral reefs

Because coral is a sedentary organism, oil spills are particularly threatening to it. Some of the most active regions in terms of oil extraction, processing and transport coincide with extensive reefs. The Arabian Gulf, the Red Sea and the Panama Canal are locations where oil and coral come into contact. The accidental release of 50 000 tonnes of crude oil from the Isla Payardi refinery, Panama in 1986 led to coral reefs, seagrass beds and mangroves being contaminated. Immediate effects were obvious, with high levels of mortality for many species, such as bivalves living among the mangrove roots and the coral itself. Oil accumulated and became incorporated within sediment held within the mangrove ecosystem. This was released gradually in the years following the spill, contaminating local ecosystems. Sediment became subject to increased wave action because less protection came from the damaged reefs and seagrass beds. From ongoing studies it seems that ecosystems can take over a quarter of a century to recover from the impact of an oil spill.

Preventing and dealing with oil spills

Methods of coping with oil spills have needed to improve significantly as the scale of tankers has increased since the 1960s. In 1967 the *Torrey Canyon* ran aground on the Seven Stones reef off Land's End, Cornwall, spilling 120 000 tonnes of crude oil. In 1978 the *Amoco Cadiz* was wrecked off the Brittany coast with the release of 230 000 tonnes of crude oil.

Prevention techniques have been achieved by improvements in ship design and methods of handling oil. Preparing for spills involves the assessment of risk and likely impact on the coastal zone, especially locations near to oil installations and along busy transport routes. The persistence of oil on different types of shore (for example, rock, sand, marsh) and the sensitivity

KEY TERMS

Advection The transfer of matter or heat horizontally in a fluid natural environment. Thus oil is moved by the velocity of sea water.

Trophic level The level at which energy in the form of food is transferred from one organism to another as part of a food chain.

of ecosystems to oil pollution are assessed and mapped. Reactions to oil spills depend on the resources of the relevant authorities. Clearly the level of resourcing is paramount, especially for the provision of machinery and manpower. A variety of techniques exist for clearing up after an oil spill. On water, floating booms are used to contain surface oil which can then be pumped into tankers. Purpose-built vessels can skim off surface oil as long as it is not too dispersed. Chemical dispersants can be effective but may be more toxic than the oil itself. The idea is to reduce concentrations of oil to promote evaporation and biological degradation by oil-consuming bacteria.

Once the oil washes up on the shore its physical removal and the disposal of bird and animal carcasses becomes the priority. Surface oil can be collected by scraping. In the example of the *Amoco Cadiz*, 35 000 French military personnel were involved for a month collecting such deposits. However, cleaning becomes very difficult once oil seeps into coastal sediments. Wholesale removal is often impractical and can have a destabilising effect on the equilibrium of sediment budgets. Techniques such as introducing oil-consuming bacteria and/or irrigating or aerating the sediment can be attempted to accelerate the breakdown of the oil.

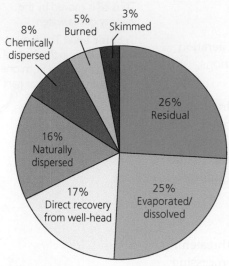

Residual = oil in ocean, washed ashore and collected or in sand and sediments
Naturally dispersed = oil currently being degraded naturally, e.g. by bacteria

▲ **Figure 6.10** Where oil from *Deepwater Horizon* has gone

In the case of *Deepwater Horizon* one question still being researched is where did all the oil go (Figure 6.10)? Not all crude (unprocessed) oil has the same chemistry. The light crude from *Deepwater Horizon* dissolves in water more readily than heavier crude and is evaporated and digested more effectively by bacteria.

Energy resources – renewables

Global energy demand continues to be met overwhelmingly by fossil fuels (coal, oil and gas) – between 75 to 80 per cent. As both population and living standards rise, demand for energy will also increase. Meeting this growing demand, as well as the pressing issue of anthropogenic climate change linked to burning fossil fuels, will increasingly turn attention to renewable energy sources. The coastal zone is the location of a variety of renewable energy resources.

Tidal energy

The rise and fall of tides brings about a flow of water which can be used to drive energy-producing machinery. Mills driven by water wheels turned by tides have existed for centuries. Around the globe, the potential for tidal energy to generate electricity is being actively investigated and some schemes initiated.

Local physical geography, such as coastline shape and tidal range, is a major factor in the location of tidal power stations. Tides are reliable, regular and predictable. Most coastlines experience two high and two low tides approximately every 24 hours (page 14). Although many suitable locations exist, any particular location needs to be close enough to where electricity demand is high enough to justify the high cost of development. (There is an economic limit to the distance electricity can be transmitted.)

The Shiwa Lake scheme in South Korea is the largest tidal power plant, having a generating capacity of 254 megawatts (MW). In Brittany, France the Rance plant can generate up to 240 MW but no other operational schemes offer more than 20 MW of power.

▲ **Figure 6.11** Tidal turbine assembled for use in the MeyGen project, Scotland

Both the Shiwa and Rance schemes use a **barrage**. Gates in the barrage open as the tide rises. At high tide the gates close, creating a lagoon. As the tide falls the stored water is released, passing through turbines and back out to sea. Several similar schemes are proposed for the UK, for example at Swansea and Cardiff and across the Severn. Issues regarding funding and environmental impacts have so far delayed any construction. Flows of water will be altered, potentially causing scouring of the estuary bed in places as well as less vertical mixing of the water. In turn this will lead to less penetration of saline water into the estuary, making the water **brackish**. Upstream of the barrage, there may be a build-up of contaminants due to less flushing by the tides and so water quality may decrease and potentially **eutrophication** might occur.

There may be a loss of certain habitats such as intertidal flats and salt marsh, with the consequent impacts on species that rely on the rhythms of the tides across these features (pages 92–4).

In the UK, the MeyGen tidal stream project in the Pentland Firth, north-east Scotland has begun generating electricity. Phase 1 is planned to have a capacity of 86 MW, while on completion the scheme is anticipated to be able to produce up to 398 MW. The scheme deploys turbines, driven by 16-metre diameter rotor blades, on the seabed to take advantage of the fast-flowing currents (5 metres/sec) in the area.

Wave energy

The potential energy in the rise and fall of water with the passing of a wave is enormous and much greater than that for tidal power. However, that very energy is a major obstacle to the exploitation of this flow resource. High wave energy, such as is found off the north-west coast of the UK, is required for the machinery to be effective. However, many devices tried so far cannot survive in these rough seas. Making the machinery more robust involves a substantial increase in size and weight, thereby decreasing the efficient conversion of movement into electricity.

KEY TERMS

Barrage A dam-like structure built across part of the coast, usually an estuary.

Brackish Water that is less saline than salt water but more salty than fresh water. It is often associated with low energy environments such as coastal lagoons landward of a spit or bar.

Eutrophication The process by which nutrient enrichment in water leads to increased primary production of algae and a reduction in oxygen levels leading to anaerobic conditions. Many organisms rely on access to oxygen and so are unable to survive.

A variety of devices are being trialled around the world. Some float on the surface as elongated cylinders or as buoys, with the passing waves moving the machinery up and down. This drives hydraulic systems which turn generators, producing electricity. Others consist of giant flaps anchored in relatively shallow water. As the flaps move back and forth as waves pass, hydraulic systems turn generators.

The technology is yet to be economically viable, but as with all power generation, viability depends on global and local energy costs.

Offshore wind

Wind energy has been exploited for thousands of years. Since the 1980s, the use of wind to generate 'green' electricity has received much attention due to technological advances. In 2016 wind power capacity made up about three per cent of global power.

Because wind speeds in the coastal zone tend to be faster than inland, coastal locations such as cliff tops and just offshore are attractive for wind turbines. Small increases in wind speed produce large increases in energy output: for example, a 15 mph wind can generate twice as much energy as the same turbine in a 12 mph flow. Ongoing research into wind turbine technology is resulting in great efficiencies in design, leading to improved energy conversion (wind flow → electricity) and falling costs. This applies especially to offshore wind farms where very large rotor blades can be installed and operated.

One of the largest offshore wind farms in the world is the London Array project with its 175 turbines climbing 80 metres from the sea surface. Its construction began in 2009 and it was completed in 2012. It covers an area of 100 km², 20 km offshore in the outer Thames estuary. Its generating capacity of 630 MW is sufficient power for nearly 0.5 million homes a year, reducing CO_2 emissions by about 925 000 tonnes per year (Figure 6.12).

▲ **Figure 6.12** London Array wind farm

Wind energy is a contested issue and leads to significant debate about the nature of a coastal place. Some object to the intrusion into a seascape with its qualities of distance to the horizon and large uninterrupted skies. However there is the view that offshore wind farms are less intrusive than those located on cliffs and exposed places just inland. Environmental concerns focus on noise, electromagnetic interference to radio and television reception, impacts on wildlife such as birds struck by rotating blades and visual impacts. Modifications to designs have reduced some of these but varying attitudes towards the presence of wind turbines in the coastal landscape remain.

Tourism and recreation

Both tourism and recreation are activities that rely on resources such as climate, landscape and water. Over the past 200 years, tourism and recreation have exploited coastal resources to become major economic and social activities. As disposable income, personal mobility and paid holiday time increase, growing numbers of people visit the seaside to enjoy its resources.

By the end of the nineteenth century, in most high income countries, seaside holidays became part of the annual rhythm of many people's lives. Sea bathing moved from being a small-scale, fashionable and therapeutic pursuit of the rich, to mass tourism. All sectors of society now participate, with some people being able to exploit coastal resources for tourism on any continent – even Antarctica.

Making coastal spaces into tourist places

As with any economic activity, tourism and recreation require resources in order to operate. The physical 'raw materials' are:

- relief, e.g. steep cliffs, low-lying land, slope of offshore gradient
- beach material, e.g. sand, pebbles
- water, e.g. nearshore water flows (rip currents/gentle currents), temperature, quality
- climate, e.g. Mediterranean, cool temperate, arctic
- ecosystem, e.g. estuary, sand dunes, coral reef.

Different combinations of physical resources result in the development of different types of tourism. Where easy access to the shore, a sand beach and a climate with a hot and dry season are found together, beach- and water-based activities flourish. Well-known examples include the resorts of the southern Californian, eastern Australian and Mediterranean coastlines.

As well as physical resources, tourism also seeks to use human resources, for example:

- cultural attractions, e.g. theatres, restaurants, nightclubs
- heritage resources, e.g. architecture, piers, preserved railways.

Some physical resources acquire cultural or heritage attributes (Figure 6.13). For example, large stretches of the coast of England and Wales are owned by the National Trust (Northumberland, north-east England) or are designated as national parks (Pembrokeshire, south-west Wales).

▲ **Figure 6.13** Physical and heritage resources used for tourism, Bamburgh Castle, Northumberland coast

Sand and gravel

The coastal zone contains vast deposits of aggregates. The construction industry makes extensive use of these materials as do many coastal management schemes.

The demand for sands and gravels has been rising across the world and in some regions the rate of use is itself increasing. The United Nations has been registering the growing urbanisation of the world's population. Some 54 per cent of people currently live in towns or cities but by 2050 some two-thirds of the anticipated global population of 9.8 billion will be urban dwellers. As urban areas expand there is an increasing strain placed on aggregates for construction, in particular sand. Concrete and glass, two essential materials in buildings, rely on sand in their manufacture. Additionally, activities such as fracking for oil and gas use large quantities of sand.

Sand is also used in land reclamation. Singapore has imported about 517 million tonnes of sand from Cambodia, Indonesia, Malaysia and Thailand. Dubai exhausted its domestic marine sand resources when it used 385 million tonnes to create the artificial set of islands, the Palm Jumeirah. The ongoing substantial construction projects in Dubai are supplied with imported sand from as far away as Australia.

It might seem as if there is an abundance and a vast quantity of sand in the world. However, sand varies greatly in its nature from one location to another. Not all sand is the same! Desert sands are little use: having been rounded by wind erosion they do not bind together well enough for construction. Until recently, most sand was extracted from land quarries and river beds. With demand outstripping supply, sands in the coastal zone are being more intensively exploited.

Impacts of sand extraction on marine ecosystems and environments

Digging or dredging for sand and gravel takes place at varying scales, from small-scale abstraction by hand digging to large-scale commercial dredging operations.

Dredge mining is carried out either from a boat fitted with a loop of buckets that scoop up sediment or with a powerful hydraulic system that sucks up sediment from the seabed.

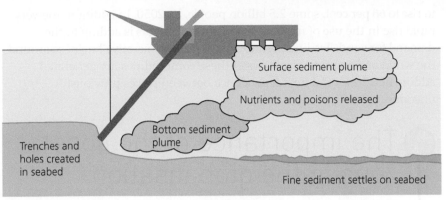

▲ Figure 6.14 Some of the effects of dredge mining

Disturbance of the seabed adversely affects the habitats of fish, invertebrates and algae, as well as physically removing bed-dwelling fauna. Food webs and chains are interrupted, which in turn impacts birds and mammals. Crucially, some dredged areas are the spawning grounds for fish and so fish stocks may decline as a consequence of sand extraction.

In the course of dredging, small-calibre material that is not required is released back into the water. This fine sediment settles back on the seabed and can suffocate filter-feeders such as mussels. Fine material can also fill crevices where shellfish such as lobsters and some species of fish live. Fish species such as herring and sand eels only spawn where the seabed comprises coarse-grained material, so the deposition of fine-grained material reduces breeding potential.

Poisons, such as heavy metals and hydrogen sulphide, may be released when dredging occurs. Plankton blooms can develop in response to the release of nutrients from the 'sea soil'. Light levels and visibility are thus reduced, affecting the ecosystem.

Depending on local circumstances, sand extraction can change the pattern of wave energy reaching the coast. Removal of sand deepens and smooths the seabed, which allows waves to pass over with less frictional interference. Therefore more wave energy can reach the shore, increasing rates of erosion and sediment transport.

Local sediment budgets can be disturbed by sediment removal. Beach and dune systems can be diminished, thereby allowing greater wave energy to impact the land behind them. Risks of coastal flooding may rise. In general, the deeper the water in which the extraction occurs, the less likely it is there will be disturbance to beach and dune systems.

Sand and gravel represent the highest volume of raw material used on Earth after water. Their use greatly exceeds their renewability by weathering and erosion. The strain on this resource comes primarily from construction, with concrete and glass reliant on sand and gravel. Currently 54 per cent of the world's population live in urban areas and this is predicted to rise to 66 per cent, some 2.5 billion people, by 2050, In addition, the very rapid rise in the use of fracking to extract oil and gas is adding to the demand for sand. Trading in sand and gravel is a lucrative business around the world and as prices rise flows of these commodities are generated, adding new links and interdependence between places previously unconnected in this way.

The importance of the coastal zone in the globalisation process

▶ *What has been the contribution of the coastal zone to globalisation?*

The nature of globalisation

At its essence, globalisation involves the growing interconnection and interdependence of people's lives. Throughout human history, individuals, tribes and states have moved out beyond their original home, often in search of additional resources. While the spatial scale of many of these searches were, by today's standards, relatively local or at most regional, the extent of some interactions was truly remarkable, especially those involving travelling by sea.

The globalisation that has been occurring since the Second World War represents a significant change to what had taken place previously:

- connections between people and places have lengthened.
- connections in terms of goods, people and data have become faster.
- there are deeper and more diverse connections affecting the daily lives of greater numbers of people.

The extent to which globalisation is experienced varies greatly among places and from one person to another (Figure 6.15).

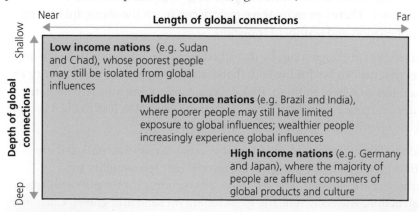

▲ **Figure 6.15** Variations in the experience of globalisation

It is important to note that virtually all examples of countries experiencing shallow and near global connections are landlocked or have a coastline with few suitable locations for major harbour development. This emphasises the role the coast and shipping plays in the globalisation process.

The development of marine navigation

The first use of the sail to power water transport was around 4500 BCE. The first urban civilisations, such as those located in Mesopotamia and the Nile and Indus Valleys, engaged in maritime trade. The eastern Mediterranean, the Persian Gulf and the coastline extending to modern-day Pakistan were corridors along which goods and people passed. The city of Dilmun, in present-day Bahrain, was a well-established **entrepôt** similar to Singapore or Rotterdam today.

As a succession of kingdoms and empires rose and fell in various parts of the world, some made almost exclusive use of transport by land but others exploited maritime movement. The Roman Empire was largely dependent on trade, much of which was carried by boat. By the first century CE, trade routes in the Indian Ocean had extended beyond the immediate coastal zone as sailors made use of the regular seasonal pattern of the monsoon winds.

In the first millennium CE, three sets of peoples were pre-eminent navigators. The Norse crossed the North Atlantic as far as Vinland, present-day north-east Canada. Arab ships sailed to the East Indies, East Africa and China. The Polynesians were perhaps the most remarkable as they colonised the far-flung islands of the Pacific Ocean.

> **KEY TERMS**
>
> **BCE** Before Common Era, indicates the number of years before the supposed year of Christ's birth.
> **Entrepôt** A centre to which goods are brought for import and export and for collection and distribution.
> **CE** Common Era, indicates the number of years since the supposed year of Christ's birth.
> **Circumnavigation** The act of sailing all the way round the world.

Distance to horizon

It is interesting to note that the distance one can see to the horizon is about 5 kilometres, assuming an eye height of 2 metres. Early sailors would have been able to climb higher up a mast, but even at 10 metres high the distance to horizon is still only just over 11 kilometres. The height of the land being looked for is also important, as the higher this is the further away one can spot it. For the Polynesians, many of the Pacific Islands are relatively low-lying mounds of land. Given the expanse of open water that many early sailors navigated, their skills at spotting signs of land such as cloud patterns and birds, as well as their courage, are very impressive.

The global world view is a relatively modern concept. During the sixteenth and seventeenth centuries, knowledge of the continental coastlines greatly increased following the 'voyages of discovery' of maritime nations such as England, the Netherlands, Portugal and Spain, the Arab traders and the Chinese. However, despite epic voyages such as Magellan's **circumnavigation** in 1519 to 1522, accurate charting of the world's oceans and coasts was not achieved until the twentieth century.

The expansion of European empires in the eighteenth and nineteenth centuries owed much to maritime power. Global power-play and ambitions continued through the twentieth century and on into the twenty-first, in particular with the USA and most recently the Chinese.

Modern globalisation and the role of ocean transport

More people and places are now connected in economic, social and political ways. Vast and highly complex networks exist, affecting so many aspects of lives around the world. Ports and ocean transport play pivotal roles in the nature of modern globalisation.

To a large extent, economic activities are key components in globalisation. Total world trade has more than trebled to 45 per cent of global GDP since the 1950s. The production and selling of a very wide range of goods affects the lives of billions of people. Whether it is the food they eat or the manufactured goods they use, increasingly there is a global element. And ocean transport is key as part of that. Australian coal is shipped to Japan, kettles made in China make their way to Europe, and grain produced in the USA and Canada arrives in the UK (Table 6.3).

Year	Oil	Bulk cargoes[1]	Dry cargo[2]	Total cargoes
1970	1442	448	676	2566
1980	1871	796	1037	3704
1990	1755	968	1285	4008
2000	2163	1288	2533	5984
2010	2752	2333	3323	8408

[1] iron ore, grain, coal, bauxite, phosphate

[2] wide range of products, e.g. textiles

▲ **Table 6.3** Growth in global seaborne trade (millions of tonnes)

Sea transport has played a vital role in a key aspect of globalisation, time–space compression.

The revolution in marine technology

New technologies have revolutionised connectivity. For large cargo ships, journey times from Shanghai are, on average, about 32 days to Rotterdam, 36 days to New York, 17 days to Sydney and 22 days to Los Angeles. Ship architecture such as hull shape and propeller design, substantial increases in engine power and improvements in navigation aids such as radar and global positioning systems based on satellites, combine to increase both the speed and reliability of voyages.

Containerisation plays a fundamental role in globalisation. Eliminating 'loose' cargo and item-by-item handling reduces costs at every stage, from the factory to the consumer. Economies of scale have contributed to falling relative freight rates (Figures 6.16 and 6.17). Loading and unloading a vessel is greatly speeded up by the use of containers. Each container has a unique barcode so that mechanised handling can operate quickly. The logistics of distribution are very efficient, with computers tracking container movements.

KEY TERMS

Time–space compression A set of processes leading to a 'shrinking world' caused by reductions in the relative distances between places due to travel times decreasing.

Containerisation The shipping of goods (by road, rail and sea) in standard-sized metal boxes. It allows efficient mechanised handling of large volumes of goods and lowers transport costs.

Economies of scale These arise when an economic activity operates at a large scale. Conveying double the number of containers at a time on a ship neither doubles the fuel costs nor does it require twice as many crew.

Pearson Edexcel

AQA

OCR

WJEC/Eduqas

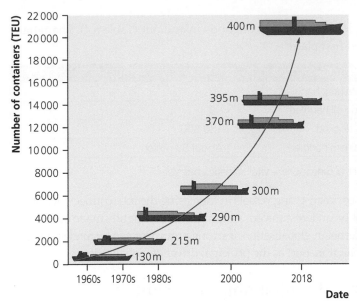

▲ **Figure 6.16** Increasing size of container ships (metres)

▲ **Figure 6.17** The world's largest container vessel, Orient Overseas Container Line (OOCL) *Hong Kong*, launched in 2016, with a capacity of some 21 400 **TEUs**

Bulk carriers of goods such as oil, mineral ores and grains have also increased in size and achieved similar scale economies. The largest oil tankers carry some 3 million barrels of oil, equivalent to 440 000 tonnes. Iron ore carriers can be nearly as big, at 400 000 tonnes.

Ocean transport is dominated by vessels carrying goods. However, similar technological changes have been affecting cruise ships. The largest of these are now up to 225 000 tonnes, 360 metres long and can carry around 6000 passengers and 2300 crew.

The impact of the increasing size in ships on ports

The coastal zone is most affected by the vast scale of vessels and their demands in terms of port facilities. The physical geography of the coast is an important influence on port location. Depth of water, tidal range and shelter are key factors. Port Valdez, Alaska is at the head of a deep fjord (page 126) and has the major advantage of being ice-free all year round. In 1989 the *Exxon Valdez* oil tanker ran aground some 40 kilometres from the port, spilling 11 million gallons of crude oil into Prince William Sound. Since then, measures taken to improve the passage of tankers into and out of the port have made it into one of the safest oil ports in the world. Natural harbours such as Sydney, San Francisco and Singapore are long-established ports but, with increasingly ambitious engineering, harbours can be developed in locations without such natural advantages. Extensive engineering such as dredging and breakwater and dock construction have allowed Europoort to develop at the mouth of the River Rhine in the Netherlands.

The scale of the larger port operations is immense. Singapore, until 2005 the world's busiest port in terms of tonnage handled, remains the largest

> **KEY TERM**
>
> **TEU** Twenty-foot equivalent unit. The dimensions of one TEU are equal to a standard shipping container: twenty feet long, eight feet wide and eight feet, six inches high.

trans-shipment location in terms of goods in and out (Table 6.4). Shanghai has now claimed first place.

Factor	Statistic
% of world's containers handled	20
% of world's crude oil handled	50
Number of ships docked	130 000
Number of containers handled	About 33 million

▲ **Table 6.4** Port of Singapore – vital statistics per year

The importance of ocean-going vessels in increasing the interconnectivity of the world should not be underestimated. Although much is rightly made of the impact of the internet in globalisation, a strong argument can be made for the revolution in ocean transport as the principal driver of globalisation.

The pattern of global shipping routes

The principal shipping routes starting and finishing from ports follow a relatively simple pattern (Figure 6.18). An east–west corridor links North America, Europe and Pacific Asia making use of the Panama and Suez Canals and passing through the strategic 'pinch-point' of the Strait of Malacca. The latter location is a narrow (just two miles at its least width), 885-kilometre-long corridor stretching between the Malay Peninsula and the Indonesian island of Sumatra. Another major route extends from Europe to eastern South America across the Atlantic Ocean. Several secondary routes add to the network, such as between Brazil and South Africa and from there across the Indian Ocean. There are, in addition, many short sea crossings which are vital to movements of people and goods. The Channel, Irish and North Seas, and among the islands of Indonesia and Greece, are examples of locations where numerous ferries operate intensive services.

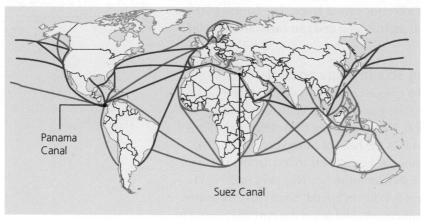

Key

—— Core route   —— Secondary route

▲ **Figure 6.18** Main global shipping routes

The role of physical factors on shipping routes

Although advances in technology have greatly improved the abilities of vessels to cross considerable expanses of water, various physical factors continue to exert significant influences. The shape of coastlines, prevailing winds, water currents, water depths, reefs, sea ice and icebergs all influence where and to some extent when particular courses are sailed by vessels. Seasonal winds such as those generated by the various **monsoons** affecting many Asian coasts can both aid and restrict sea journeys. The annual expansion and contraction of sea ice across the Arctic has been one of the strongest physical influences on shipping across the centuries. Not only have routes been actually shut off or opened, the risks posed by icebergs are not to be underestimated. The impact of global warming in reducing the thickness and extent of Arctic sea ice is beginning to influence shipping routes. Both the Northwest Passage across the north of Canada and the Northern sea route across the north of Siberia are being explored to assess their commercial potential in moving goods.

The two pre-eminent canals as regards ocean-going vessels, the Suez and Panama, have had far-reaching influences on shipping routes. The former, opened in 1869, saves the journey via the cape of Good Hope, South Africa, while the latter, opened in 1914, relieved ships of the treacherous passage around Cape Horn, South America. Currently neither can take the larger ships, but the capacities of both canals are being upgraded to allow more and larger ships to pass through.

The direction and type of trade across oceans

From the coastlines of all the continents except Antarctica, vessels depart and arrive carrying a very diverse set of cargoes. The terms merchandise and commodities are frequently used when referring to goods. Both describe any goods, materials and products, but not services such as banking and insurance. A wide variety of data is collected by a range of organisations, such as the World Trade Organization (WTO), the United Nations Conference on Trade and Development (UNCTAD), the International Monetary Fund (IMF) and the Organisation for Economic Co-operation and Development (OECD). There are also groups that represent a particular commodity such as the Organization of the Petroleum Exporting Countries (OPEC) for oil and the International Coffee Organization (ICO) for coffee.

Among the diversity of trade, one feature emerges very strongly: the uneven pattern of trade. Wherever one looks within trade statistics, trade is dominated by the more advanced and rapidly emerging economies. These nations possess the economic, political and social resources to sustain their controlling roles in world trade. The **terms of trade** are much weaker for the least developed countries as they have more limited access to global markets. Often a key factor in that restriction is a poor trading infrastructure such as lack of port capacity. This might be because the approach channel is too shallow for cargo vessels and the country does not have dredgers to maintain a deep water channel. The actual port may lack berths and cargo-handling facilities such as container cranes and then may not have the inland distribution networks of rail or road.

 KEY TERMS

Monsoon A seasonal reversal of wind direction. The best known monsoon occurs in India and South-east Asia.

Terms of trade The value of a country's exports relative to that of its imports. This is measured as: average price of exports/average price of imports x 100. If export prices rise relative to import prices, there is an improvement in that country's terms of trade.

Primary goods Economic activities that produce food, fuel and raw materials. These are unprocessed and so have only a little added value.

Secondary goods Economic activities that process/manufacture goods thereby adding value to whatever is being produced, from an armchair to a X-ray scanner.

While **primary goods** exports generate some income, they have less added value than **secondary goods**. The pattern of the ten leading exporters and importers as seen in agricultural products (an example of primary goods) and manufactured goods reveals some interesting contrasts (Tables 6.5 and 6.6).

Top ten exporters of agricultural products			Top ten importers of agricultural products		
Country	Value (billion US$)	% share of world trade	Country	Value (billion US$)	% share of world trade
EU	585	37.1	EU	590	35.0
USA	163	10.4	China	160	9.5
Brazil	80	5.1	USA	149	8.8
China	73	4.6	Japan	74	4.4
Canada	63	4.0	Canada	38	2.3
Indonesia	39	2.5	South Korea	33	2.0
Australia	36	2.3	Russian Federation	28	1.6
Thailand	36	2.3	Mexico	28	1.6
Argentina	35	2.2	India	28	1.6
India	35	2.2	Hong Kong (China)	27	1.1

▲ **Table 6.5** The top ten exporters and importers of agricultural products 2015 (US$ billion). Source: WTO International Trade Statistics, 2015

Top ten exporters of manufactured goods			Top ten importers of manufactured goods		
Country	Value (billion US$)	% share of world trade	Country	Value (billion US$)	% share of world trade
EU	4239	36.6	EU	3812	32.9
China	2153	18.6	USA	1808	15.6
USA	1126	8.7	China	1084	9.4
Japan	545	4.7	Hong Kong (China)	506	4.1
South Korea	470	4.1	Japan	372	3.2
Hong Kong (China)	442	4.0	Canada	323	2.8
Mexico	312	2.7	Mexico	320	2.8
Singapore	266	2.3	South Korea	269	2.3
Taiwan	240	2.1	Singapore	206	1.8
Canada	208	1.8	India	187	1.6

▲ **Table 6.6** The top ten exporters and importers of manufactured goods 2015 (US$ billion). Source: WTO International Trade Statistics, 2015

Economically advanced countries dominate all four lists, with a few rapidly emerging countries in addition. Every one of these countries has an extensive coastline with at least one large port capable of handling the scale of vessels and cargoes demanded by the trade it is engaged in.

Submarine cables – connections along the seabed

Although out of sight and, away from the coastal zone, at substantial water depths, the seabed is vitally important to globalisation. This is because a network of cables of different types criss-crosses the oceans carrying communications and electricity.

Submarine communication cables

During the second half of the nineteenth century, submarine cables were laid across ever-increasing distances. Europe and North America were first linked by cable in 1866. By the 1950s a global network of telephone cables had been installed with modern networks now using fibre optic links. The internet could not function as it does without these underwater cables. It is an indication of the growing significance of countries in Asia Pacific that the greatest level of cable-laying activity in the past two decades has been in the Pacific (Figure 6.19). Cable laying is also taking advantage of the opening up of the Arctic with links between London and Tokyo being developed.

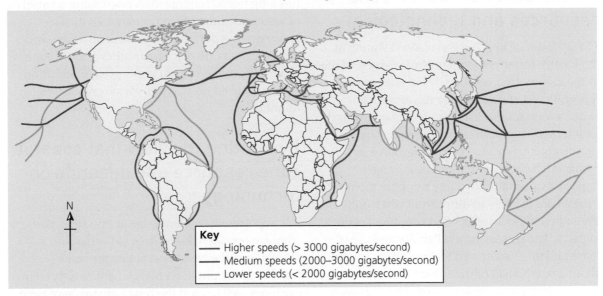

Key
— Higher speeds (> 3000 gigabytes/second)
— Medium speeds (2000–3000 gigabytes/second)
— Lower speeds (< 2000 gigabytes/second)

▲ **Figure 6.19** Main global submarine communication cable network

Submarine power cables

Submarine power cables were first laid in the 1950s. Since then, some 8000 kilometres of cable has been installed, the vast majority within Europe. Connecting islands to mainland supplies of electricity is a major consideration but there are also several cables linking countries. There are existing links between, for example, Norway and Germany and Denmark, between Sweden and Finland, between Sweden and Germany, between Eire and the UK and between the UK and France. Surplus energy can be moved along cables to locations with a deficit. Variations in weather conditions in different countries can create either surplus or deficit situations which energy flows along cables can help even out. Different levels of demand and supply can also arise when two countries have different public holidays.

The nature of the seabed's topography offers considerable challenges to engineers, even across relatively short corridors such as the North Sea and the Channel. Obstacles such as a tectonic plate boundary, mid-ocean ridges or ocean trenches associated with subduction zones, mean that cable laying cannot be accomplished everywhere.

 Evaluating the issue

▶ *To what extent does a society's use of coastal zone resources depend on its access to technology?*

Identifying possible contexts for the study of coastal zone resources and technologies

The widespread and large-scale exploitation of the Earth's resources is a significant aspect of life in the twenty-first century. There is a growing recognition that how we use resources has various consequences in the short, medium and long terms. These different time scales are significant as they affect how a resource is managed. If the resource can be adversely impacted in a matter of a few weeks or months then its use is likely to differ from when only use over a period of decades will bring about serious impacts. The increasing availability and use of more and more advanced technology is bringing about a reappraisal of the value that resources have for human activities. Nowhere is this seen more clearly than in the use of energy resources located in the coastal zone.

There is no one single agreed definition of a resource but in general it is something that satisfies human wants and needs. In addition, the degree of renewability of a resource is a factor to consider. Some resources are infinitely renewable (solar and wind) while others can be renewable depending on levels of use and management, while yet others are exhaustible, such as fossil fuels.

Resources can be identified at a variety of scales, from global down to very local. The coastal zone has long been recognised as possessing a variety of resources valuable to human activities. However, the definition of resources is dynamic, varying from one place to another and through time. Within the coastal zone, greater attention is now focused on what its various resources might offer.

Evaluating the view that coastal zone resource use depends on technology

It is important to acknowledge at the outset that what are considered resources today has not always been the case. Most substances and features of the physical environment have been in existence well beyond the time humans have been around, things such as sand and waves. The important point is not their existence but rather what function they perform in human society. For example, although coal had been available for centuries, albeit not in large quantities, it was the harnessing of steam power, fuelled by coal, which underpinned the profound changes associated with what is termed the 'industrial revolution'. In this context, the technological changes of the late eighteenth and nineteenth centuries went hand in hand with redefining coal as the pre-eminent fuel. What establishes whether something is a resource or not has an important bearing on the relationship between a society and its natural environment.

Because the idea of a resource is dynamic and closely linked to the evolution of human societies, it follows that technological change can have significant impacts on resource definition and therefore use. Throughout history as new technologies have emerged, particular resources have either become valuable or redundant. Some substances remain as resources but the ways in which we use them alters.

The rather simple classification of resources into renewable and non-renewable can provide a starting point. The coastal zone contains many resources of both categories. Within the coastal zone there are a variety of resources that can be exploited for energy. Some of these are renewable while others are not. The renewable or flow resources include wind, wave and tidal energy. These three types have been recognised by human societies across the centuries but varying use has been made of them.

The power of the tides has been recognised for centuries. Medieval tidal mills were set up at locations with a significant tidal range, such as the British Isles. However, the turning of a water wheel does not generate that much power, although at the time such technology made a significant positive impact on the people benefiting from its use, for example by the grinding of their wheat into flour.

Recent technological advances have made possible generation of considerable quantities of electrical power from tidal flows of water through coastal barrages such as the Shiwa scheme in South Korea. As the search intensifies for more electricity-generating capacity from renewable sources, it may be that barrage schemes become more widespread.

Wave energy has long been recognised as having considerable potential, but the various technologies tried so far have not been successful. However, some projects are offering more commercially viable prospects as operational designs improve. In the short term, wave energy is likely to be more experimental but if the technology can be proven to be robust and reliable, then it may become commercially viable.

Average wind speeds around the coast are higher than locations inland. The sea surface offers limited resistance and friction to moving air. Technological advances now allow giant wind turbines to be located in vast 'farms' offshore, such as the London Array, with the electricity transmitted onshore.

Evaluating the view that coastal zone resource use does not always depend on technology

The power of the waves has long been recognised and for many centuries it was probably seen as a threat rather than an advantage. Until relatively recently, high levels of wave energy brought disaster to vessels as their technological capabilities were limited. Prediction and forecasting were rudimentary and, with a reliance on sails and perhaps oars for power, sailors had few options. However, with advances in marine engines and steering systems, as well as much more accurate and reliable weather forecasting, navigation has become less hazardous. On the other hand, it is the case that even the most modern vessels continue to shelter in bays when major storms pass through an area. Rough seas continue to represent serious threats to vessels despite significant advances in the technologies of ship design and construction. Ferries around the world have their services interrupted by adverse weather and sea conditions such as across the Channel and North Sea.

For centuries, salt has been obtained by evaporating seawater in shallow lagoons known as salt pans.

The process is powered by the sun and is something that continues to produce the valuable resource of salt today without a significant input of technology.

The value of the coast as a location for various types of recreation and tourism has been aided by technology such as advances in transport allowing more people to visit the coast. Increased personal mobility as a result of the development of rail, road and air transport technologies has led to mass tourism in places such as along stretches of the Mediterranean and Florida coastlines. It is interesting how the growth of one technology, air transport, has contributed to a decline in the use of some coastal zone resources. Resorts such as Minehead, Margate and Skegness that grew due to developments in rail networks declined once people were able to fly further to use coastal zone resources such as warmer and sunnier climates.

However, activities such as walking, painting, watching wildlife and simply playing on the beach require little by way of technology. The coastal resources being used do not need technology to be available.

Fish have long been used by humans as a source of food and products such as oils and fertilisers. While advances in technologies have allowed more intensive use of fish stocks, growing concern over the sustainable use of some species has led to social and political power being used to limit their exploitation. The collapse in fish numbers in some locations has resulted in limits and bans on their use. To some extent, it is the very successful application of modern technologies, such as sonar and machinery capable of handling the vast nets some boats deploy, that has led to overfishing and the subsequent reduction in use of fish.

Political and social influences can also play prominent roles in outweighing technological know-how in the use of energy resources in the coastal zone. The construction of a barrage is a highly contested project with strongly-held views both for and against. The technology is available to build more of these, but environmental concerns, including the value of ecosystem services of features such as salt marsh, cause concern. Perhaps of greater significance is the cost of energy. If there was an overwhelming economic case in favour of energy being generated from barrages, then it is highly likely that more would be being built. However, given the volatility of energy prices and the difficulty in accurate price forecasting, the level of capital investment and the lead time required for construction mean that tidal barrages are not being given priority.

Arriving at an evidenced conclusion

Both societies and technology are dynamic and as both alter the use that can be made of coastal zone resources changes. It is clear that over the past two hundred years, significant advances in technology have allowed more intensive resource use over a greater spatial extent. Fishing for example is carried out by large ocean-going vessels travelling long distances from their home ports. People travel intercontinentally to use coastal resources for tourism such as scuba diving on coral reefs. The potential for the coastal zone to supply energy has begun to be realised, in part due to advances in technology, oil and gas drilling and offshore wind farms, for example.

On the other hand, technology has not allowed complete control over the coastal zone and thereby resource use is limited. Wave energy is not yet a commercial reality and neither is completely unhindered navigation. Social and political factors can outweigh technological capacity, thereby restricting a resource's use. As global population heads well over 7 billion and on towards 9–10 billion by 2050, increasing attention will be given to the use of coastal zone resources. This is likely to result in growing tension as to how and where resources can be used. Food, energy and space are possible to obtain in the coastal zone, but the extent to which such use goes ahead is more likely to rest with human attitudes rather than the level of technology.

Chapter summary

✔ The coastal zone has, for millennia, attracted human activities with a high proportion of the world's population living relatively close to a coastline. There are, however, significant latitudinal variations in the density of coastal populations, with the mid- and low latitudes more densely settled compared to the high latitudes.

✔ Many economic activities benefit from locating in the coastal zone, especially if involved in the import and/or export of goods. The best port locations are where physical and human factors coincide, such as deep water and inland resources. Opportunities for land reclamation have been taken where demand for space and/or risks from flooding are high.

✔ The coastal zone possesses a variety of resources humans can use. Prominent among these are energy resources. Hydrocarbon and renewable energy resources have been exploited with both sets being contested activities, having both advantages and disadvantages.

✔ The process of globalisation has been strongly influenced by the coastal zone. At the centre of the growing flows of goods, services and people have been developments in marine transport. There is a clear pattern of global shipping routes influenced by both physical and human factors, with the directions and types of trades reflecting and sustaining the uneven development around the globe. In addition, an increasing network of submarine cables provides links along which information flows, further enhancing globalisation.

✔ Tourism and recreation exploits physical, cultural and heritage resources located at the coast, such as beaches, climate and aspects of the built environment. Flows of people at a variety of scales seek to exploit these resources, with international tourism being a strong growth area.

Refresher questions

1 Describe the global pattern of coastal population.

2 Suggest reasons why many metropolitan regions have coastal locations.

3 What is meant by the terms 'resource', 'natural capital' and 'human capital'?

4 Outline why resource sustainability is a complex idea.

5 Explain how land reclamation in the coastal zone brings advantages for human activities.

6 Describe the role of technology in exploiting energy resources in the coastal zone.

7 Describe and explain how each of the following categories of resources can be used in tourism and recreation in the coastal zone: physical, cultural, heritage.

8 Explain the role of sea transport in time–space compression.

9 Suggest reasons for the global pattern of shipping routes.

Discussion activities

1 Using Table 6.1, discuss reasons for the contrasting pressures different countries are experiencing as a consequence of the proportion and number of their inhabitants living close to sea level. Assess what responses the countries are likely to be able to take, given where they are along the development continuum including economic, technological and educational resources.

2 Discuss the role physical geography can play in shaping the characteristics of different places such as a major resort, port or small village. To what extent might human activities have changed any physical features and how significant might be these in making a place?

3 Consider how you have benefited from globalisation – think about aspects of your life such as your food, clothes, household goods (TV, fridge, furniture, etc.), music, holiday destinations. To what extent have these benefits been due to coastal features such as harbours?

4 Divide up into groups of two/three individuals. Half the groups should construct the argument that exploiting coastal resources is essential in order to promote economic and social sustainability. The other half should take the view that environment sustainability should come first. Debate the issues. Draw up detailed guidelines for how the coastal zone should be used given your conclusion based on an Ordnance Survey map of a 20–30 kilometre stretch of coastline. Use resources such as Google Maps, Environment Agency information and the local authority strategic plans to inform your discussions.

FIELDWORK FOCUS

A *Assessing the characteristics of a coastal resort that give rise to its 'sense of place' can make use of a variety of sources including both formal and informal representations.* A starting point for a UK location is the Ordnance Survey 1:25 000 map to assess the physical characteristics. Most resorts have material on local government websites seeking to promote their attractions. Analysing commercial postcards and images posted online by visitors can offer insights into informal representations. Consulting the formal representations offered in the Census as regards age structure and occupation can add to the place profile.

B *Exploring attitudes towards the siting of a wind farm on coastal cliff tops or offshore in the coastal zone could make the basis of an A-level investigation.* Interviews and questionnaire surveys are possible data collection techniques to use, as well as visual and perhaps noise assessments. Sampling could cover various groups such as different ages, length of residence in the locality and visitors.

Further reading

Your course textbook has material on topics such as globalisation and global trade.

Crown Estate (2015) *Aggregate Dredging and the Suffolk Coastline - a Regional Perspective of Marine Sand and Gravel Off the Suffolk Coast Since the Ice Age*. London: Crown Estate

Daniels, P., Bradshaw, M., Shaw, D., Sidaway, J., (eds) (2012) *An Introduction to Human Geography* (4th edition), Chapters 14, 16 and 19. Harlow: Pearson Education Limited

Department for Transport (2017) *Transport Infrastructure for our Global Future – A Study of England's Port Connectivity.* London: Department for Transport

Murawski, S.A., Fleeger, J.W., Patterson, W.F., Hu, C.M., Daly, K., Romero, I., Toro-Farmer, G.A. (2016) 'How Did the Deepwater Horizon Oil Spill Affect Coastal and Continental Shelf Ecosystems of the Gulf of Mexico?', *Oceanography*, 29(3), pp.160–73

Neill, S.P., Vogler, A., Goward-Brown, A.J., Baston, S., Lewis, M.J., Gillibrand, P.A., Waldman, S., Woolf, D.K. (2017) 'The wave and tidal resource of Scotland', *Renewable Energy*, 114, pp.3–17

Neumann, B., Vafeidis, A.T., Zimmermann, J., Nicholls, R.J. (2015) 'Future coastal population growth and exposure to sea-level rise and coastal flooding – a global assessment', *PLoS ONE* 10(3): e0118571

World Energy Council (2016) *World Energy Resources: Marine Energy – 2016.* World Energy Council

CHAPTER 7

Coastal risk, resilience and management

While coastlines offer many advantages to human activities, they are also locations of significant risk. Not only are the numbers of people living along coastlines rising, but also the nature and severity of risks appear to be increasing. Much research is being undertaken to understand the risks with the aim of devising strategies to manage them. A wide variety of methods are being assessed to increase the resilience and sustainability of coastlines and the communities living along them. This chapter:

- analyses risks to human activities along coastlines
- investigates ways of managing for resilience, including hard and soft engineering and the management of coastal ecosystems
- explores different approaches to planning for coastal change
- evaluates the advantages and disadvantages of hard coastal defences.

KEY CONCEPTS

Adaptation How individuals, households and communities respond to and cope with changed circumstances. Increased risks from coastal flooding might encourage restrictions to be placed on new developments or even 'setting back' settlements on lowland coasts.

Mitigation Action which is taken to lessen the impact of environmental events. A wide variety of mitigation responses are practised in the coastal zone, such as hard engineering (sea walls and groynes) and soft engineering (dune stabilisation and mangrove restoration), in order to reduce impacts of coastal erosion and flooding.

Resilience The ability of individuals, households and communities to resist, absorb and recover from the effects of shocks or stresses, which can be economic, social or political. Stresses are increasing for many low-lying locations due to rising sea level for example, with the associated impact of increased probability of flooding. Shocks such as tsunami can provide a severe test of resilience. The concept is also applied to ecosystems facing stress and shock (termed ecological resilience).

Risk The probability of a range of possible outcomes resulting from specific events. Risk depends on the type and nature of the environmental event, the probability of its occurrence, its magnitude and the relative vulnerability of the people and environment that might be affected. Risks in the coastal zone include cliff collapse, flooding from storm surge and coral bleaching.

Sustainability At its most basic this refers to the idea of the use of something that can continue in ways that respect the needs of future generations. A key aim is to ensure the integrity, productivity and health of environmental and economic systems. The intention is to achieve a balance between supply and demand without having adverse impacts on the physical and economic environment. In the context of climate change, the sustainability of coastal environments and communities is a serious challenge. The management of risks faces the issue of cost versus benefit, for example in deciding whether to defend a stretch of coastline or to let natural forces take their course.

① Risks to human activities along coastlines

▶ *What are the main risks to human activities in the coastal zone?*

The relationship between humans and the coastal zone has often been uneasy. Despite the potential for food supply and trade, for centuries the sea has been respected if not feared. Informal representations of coastlines frequently highlighted the threats people perceived the coast and its adjacent sea to hold for them, most vividly the risk of loss of life in major storms.

There were, however, opportunities which were exploited by some such as fishing and abstracting salt. Through technological advances, the sea and the coast became more accessible to people so that resources in the coastal zone were available. Settlement, industry, trade, energy, food, recreation and tourism have identified the coastal zone as rich in possibilities for development. But, because the coast is where the land, atmosphere and the sea meet, the potential for episodes of high energy to occur is significant.

It is extremely unlikely that the rate of climate-driven sea level rise can be mitigated in the short- to medium-term time scales, that is up to at least the end of this century. It has been estimated that annual coastal flood damage costs alone could potentially reach somewhere between 0.3 and 9.3 per cent of global Gross Domestic Product (GDP) by 2100. Risks to human activities focus on flooding (pages 128 and 137) and erosion.

The threat of cliff erosion

Many coastlines are relatively fragile in geomorphological terms. High energy levels from winds and waves, currents and tides mean that flows of materials can be disrupted. Some geologies are relatively soft and susceptible both to marine and sub-aerial breakdown, promoting mass movements. Where both marine and sub-aerial forces act on a cliff line, cliff recession can be a significant risk.

Cliff systems (page 54) highlight the interactions among a variety of factors that produce a particular type or shape of slope. In terms of cliff recession, factors influencing erosion at the base of the cliff are usually vital (Figure 7.1).

Pearson Edexcel

AQA

OCR

WJEC/Eduqas

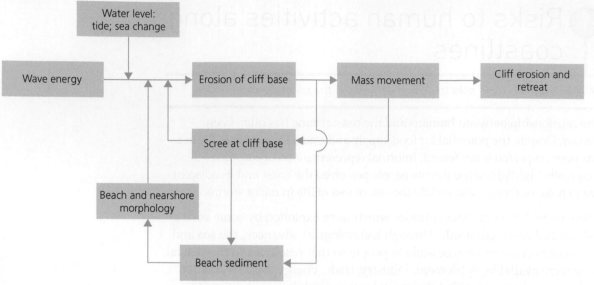

▲ **Figure 7.1** Interactions among factors controlling cliff base erosion

As waves erode the cliff base, sometimes called the toe of a cliff, the slope angle and shear stress tend to increase (page 33). The quantity of scree or talus produced and how quickly this is broken down and transported away are key influences on recession rates. Beach sediment can play two roles: one acting to promote erosion by being the material used to abrade the cliff face, the other acting to absorb wave energy and offer some protection to the cliff base.

Measuring the rate of cliff recession

The rate at which cliffs recede is very significant to human activities. Assessments need to be carried out that consider the level of threat to homes, other buildings and infrastructure such as roads and railways. The level of threat is nearly always assessed in terms of distance from the cliff edge and the average rate of cliff recession.

Various techniques are used to record the rate of cliff recession. Printed maps across as long a time period as possible and aerial photographs have been widely used. Recently, airborne and terrestrial LiDAR (Light Detection and Ranging) techniques and repeat dGPS (an enhanced GPS providing improved location accuracy) are being used to measure changes in cliffs with greater reliability and accuracy.

One issue is that average rates of cliff recession can be misleading. The further back in time, very often the less reliable and accurate the data source, old maps for example. However, what average rates disguise, as do all averages, are the extremes. Much change in the coastal zones occurs not gradually, but in sudden high energy events (such as severe storms or tsunami) that cause substantial alteration to landforms and indeed landscapes. A cliff made of mudstone near Santa Cruz, California was cut

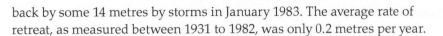

back by some 14 metres by storms in January 1983. The average rate of retreat, as measured between 1931 to 1982, was only 0.2 metres per year.

Cliff recession can also vary significantly along relatively short stretches of coastline. Factors such as variations in lithology, aspect of the cliff and the nature of the offshore gradient, influence rates of cliff recession. Flows of sediment can have a significant influence, especially where management employs structures such as groynes.

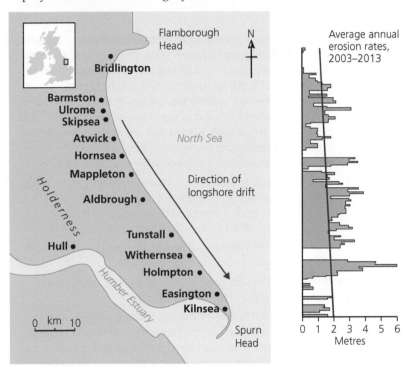

▲ **Figure 7.2** Variations in rates of coastal recession, Holderness, eastern England, 2003–13

The combination of variations in rates of erosion through time and space present additional challenges to assessments of risk and then any management considered necessary. It is clear from studies around the world that historical rates of cliff recession do not always offer a reliable guide to future rates. Equilibrium and feedback with the cliff system is complicated and with unknown factors such as future sea level rises to take into account, predictions of where, when and to what extent a particular cliff line will fail cannot be that precise.

Cliff recession on 'soft' rocks

Although all cliffs are eroded, of greatest concern to human activities is the erosion of cliffs made of relatively 'soft' rocks such as clays and weak sandstones. Granites, basalts and strong limestones generally result in far fewer issues for people.

Areas in the UK that have a lowland coast made up of 'soft' sediments are most at risk from coastal erosion. The coastlines of east Yorkshire, East Anglia and along the Thames Estuary record significant rates of recession (Figure 7.2). Current estimates have 113 000 residential and 9000 commercial properties and 5000 hectares of agricultural land within areas potentially at risk from coastal erosion.

CONTEMPORARY CASE STUDY: SOUTHERN CALIFORNIA'S RETREATING COASTLINE

The cliffs of southern California are notoriously unstable. The cliff systems include components such as:

- tectonically active fault zones, e.g. San Andreas Fault
- geologically young rocks, many from the Quaternary; 'weak' rocks, e.g. shales, siltstone, poorly cemented sandstones
- majority of the cliffs low-relief, 10 to 30 metres high
- shore platforms not particularly extensive with narrow beaches of sand or shingle/pebbles
- average tidal range about 2 metres – can reach 3 metres
- swell waves arrive from North Pacific in winter and from South Pacific in summer
- sub-aerial processes influenced by dry summers and occasionally wet winters; most rainfall November to March; heavy rain causes mass movements
- strong **El Niño** events result in higher winter rainfall, strong onshore winds, increased wave height and higher sea levels.

▲ **Figure 7.3** Intensely developed cliff line undergoing active erosion, Encinitas, southern California

California's coastal zone is a draw for millions of people. As somewhere to live, it offers access to beaches and the sea and a Mediterranean climate with cooling breezes. Employment opportunities are available from a wide variety of manufacturing and service industries. Population densities are high, especially in conurbations such as Los Angeles and San Diego. Long Beach, part of the Los Angeles conurbation, has a population density of 3600 per km^2.

Coastal defences, such as hard engineering, have been deployed to offer protection in some locations. A recent study investigated cliff recession in the first decade of this century. Its findings indicated that:

- cliffs with no base protection retreated about three times more than protected cliffs
- cliffs with beaches retreated about half as much again as cliffs with no beach.

The first finding is expected because of the absorption of wave energy by the protection method. However, the second appears counter-intuitive as it is expected that a beach will absorb wave energy. It may be that beaches form in areas of particularly soft rock that is highly susceptible to sub-aerial processes. This rock therefore crumbles readily to produce the beach. Another factor may be that the beach sediment is used to abrade the cliff and so cause a faster rate of retreat.

The issue of what to do as regards coastal management is becoming increasingly contentious and is causing much tension among the players involved. At Del Mar, a coastal community between Los Angeles and San Diego, the California Coastal Commission has raised the prospect of adopting a strategy of managed retreat. Local homeowners are adamant that the cliff line along which their properties are located should be defended from wave energy eroding the cliffs. Their intention is to have large-scale beach nourishment widen the shore which would act to absorb more wave energy. However, beach nourishment has had very mixed effects at other locations in California as well as the cost–benefit analyses not always indicating that it is worthwhile.

With rising sea level and possible increases in the frequency and/or intensity of El Niño events as a result of global warming and with the high risk of earthquake activity, California's coast may not be sustainable as home to so many people. In the meantime, much research is continuing along with planning and execution of coastal management techniques.

 KEY TERM

El Niño The name given to an unusually warm area of surface water around the equator in the eastern Pacific. They occur on average every three years, causing major disruption to the world's weather and marine food chains off the coast of South America.

② Managing coastlines for resilience with hard and soft engineering

▶ *What management techniques are used to increase a coastal location's resilience to erosion?*

The ability to cope with risks reflects the resilience that a country, community, household or individual has. Humans have sought to reduce or eliminate risks that occur in the coastal zone, such as flooding or cliff failure, by a variety of methods.

Hard engineering

Over most of the past 150 years, resilience has been sought by the construction of timber, concrete or rock structures, often on a large scale. The aim was to so strengthen the coastline with the result that wave energy in particular could be resisted.

Preventing marine erosion

Many types of **sea walls** have been constructed with variations in materials, their profiles and whether they are permeable or impermeable. They tend to be constructed directly on to coastal landforms and aim to prevent both erosion of the coast and seawater flooding inland (Figure 7.4).

 KEY TERM

Sea walls Physical structures aimed at preventing erosion and flooding.

(a) Vertical

Vertical wall, e.g. granite blocks; reflects some
energy but subject to the full force of waves

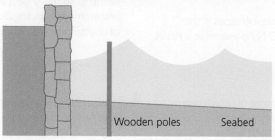

Wooden poles   Seabed

(b) Curved

Curved concrete wall

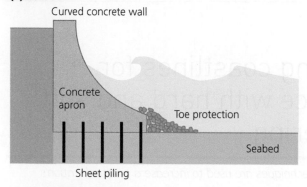

Concrete
apron

Toe protection

Seabed

Sheet piling

(c) Curved and stepped

Curved and overhanging
concrete wall

Concrete steps

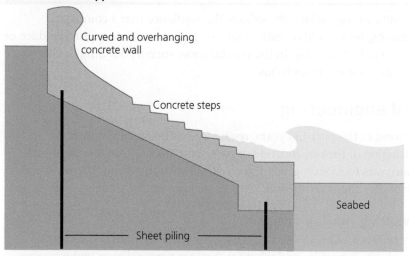

Seabed

Sheet piling

▲ **Figure 7.4** Sea wall design and effects on wave energy

Early vertical walls tended to suffer from significant turbulence at their
bases resulting in scouring away of sediment, undermining of the wall and
its eventual collapse. Toe protection is now a key part of sea wall design.

Particular issues arise where a sea wall ends. Waves refract round the end
of the wall leading to erosion of the shore, called outflanking (Figure 7.5).

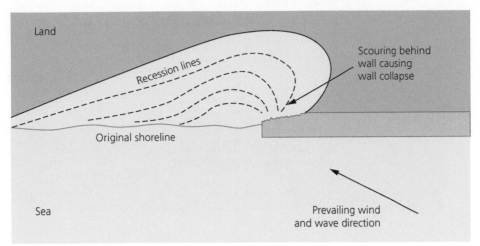

◀ **Figure 7.5** Erosion caused by outflanking at the end of a sea wall

Gabions have been increasingly deployed where less wave energy is present. They can also offer a more cost-effective method when the level of protection required is not that high, for example where land values are lower. They are effective at dissipating wave energy and when filled with local stones/rocks offer a more sympathetic 'feel' to the defence, compared to a concrete wall for example.

Revetments are made from a number of materials, with wood, rock and concrete commonly used. They are open structures so that as advancing waves break, wave energy is dissipated and less scouring occurs. Sediment can accumulate within and behind the revetment, which can also encourage vegetation to establish.

The terms 'rock armour' and 'rip-rap' are widely used when large rock boulders form a permeable barrier. Granite or basalt rocks are often used because of their resistance to both marine and sub-aerial attack. Some locations where such geologies are unavailable use precast concrete shapes, especially to protect shore areas where an industrial plant, such as an oil refinery, or a container dock is located (pages 136–7). As with revetments, wave energy is dissipated among the boulders.

Embankments or **dykes** were one of the first examples of hard engineering to be used – there is evidence of their use in Roman times. They tend to be located just above the mean spring high tide level and are used when there is an area of low-lying land to protect from flooding. Many embankments were built just landward of a salt marsh in estuary environments so that land could be reclaimed.

One issue that is becoming apparent with embankments is that of **coastal squeeze**. Due to sea level rise the width of the salt marsh reduces as more of the marsh is covered by an increasing depth of water. There is a loss of habitat and a reduction in the ecosystem services salt marshes can offer (Figure 7.7).

 KEY TERM

Gabions Wire-framed cubes filled with pebbles/rocks. They are usually filled in situ using local materials.

▲ **Figure 7.6** Gabion wall, Hengistbury Head, Dorset

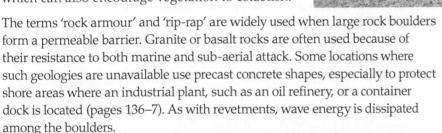 **KEY TERMS**

Revetment A generic term used when a protective layer is placed on a sloping surface.

Embankments/dykes Tend to be relatively low walls built of natural material such as clay.

Coastal squeeze A salt marsh is 'squeezed' because the embankment prevents a natural migration landwards of the marsh as sea level rises.

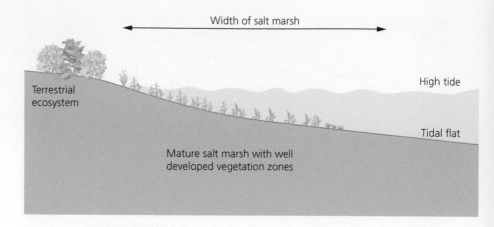

Width of salt marsh

Terrestrial
ecosystem

High tide

Tidal flat

Mature salt marsh with well
developed vegetation zones

Reduced width of salt marsh

Reclaimed land, often
agro-ecosystem

High tide

Embankment

Tidal flat

Immature salt marsh
with only lower marsh
community

▶ **Figure 7.7** The
effects of coastal
squeeze

🔑 **KEY TERM**

Breakwaters Barriers built
to protect an area from wave
energy.

Breakwaters have long been used to enclose areas of sea to provide
sheltered locations for shipping. When built for this purpose they tend to be
made of concrete. They can also be deployed to protect stretches of the
coast at risk from high wave energy. These artificial reefs tend to be made of
rock boulders or concrete shapes which absorb wave energy (Figure 7.8).

▲ **Figure 7.8** Offshore rock breakwaters and rock groyne, Sidmouth, Devon

Preventing sub-aerial weathering and mass movement

As well as countering the effects of waves on the base of cliffs, management
also attempts to deal with processes operating on the upper parts of cliffs.

A key component of the cliff system that can greatly increase rates of sub-aerial weathering and the downslope movement of material is water. Water lubricates the slope material, reducing friction between sediment particles, thereby decreasing shear strength and increasing shear stress. Solution and hydrolysis weathering, for example, can weaken rock making the slope more likely to move downslope (page 32).

Drainage of cliffs is a technique widely used to remove water as quickly as possible from a cliff. Permeable pipes can be inserted in a cliff to collect and channel water out of the cliff face. Impermeable barriers such as metal sheets are sometimes inserted, again with the purpose of preventing the cliff becoming saturated.

At some locations mass movement in the form of rockfall is a prime concern. Various techniques are applied in such situations, such as pinning the rock face by means of steel rods inserted into the cliff face. In other locations, substantial netting is secured over the cliff face to catch any material breaking free from the cliff (Figure 7.9).

▲ **Figure 7.9** Steel netting to prevent rock fall on to railways tracks, Teignmouth, Devon

Promoting sediment build-up

The capacity of sediment to absorb and dissipate wave energy is very valuable as regards modifying potential hazard events such as wave erosion and flooding. As early as the seventeenth century there are records of hard structures being used to encourage sediment accumulation. Groynes are the most common technique designed to slow longshore movement of sediment and to allow a beach to build up its height and width (Figure 7.10). Japan has deployed about 10 000 groynes as part of its coastal defences along some 32 000 kilometres of coastline.

◀ **Figure 7.10** Wooden groynes being renovated as part of the management of Dawlish Warren, Devon

Groynes are constructed from a variety of materials: wood, concrete or boulders (rip-rap). Groynes are rarely used singly but tend to be placed in a groyne field along a beach. A critical aspect of their successful use is the ratio between groyne length and spacing (Figure 7.11). On sandy beaches, a ratio of 1:4 seems to be the most effective, while on gravel and shingle beaches 1:2 is considered optimal.

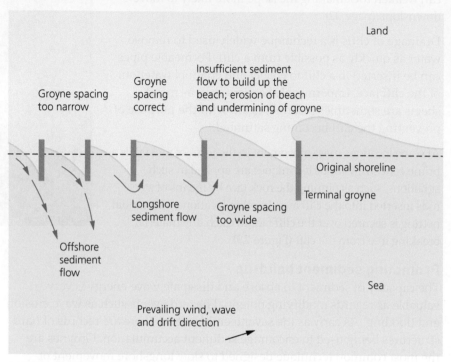

▲ Figure 7.11 The effects of groyne spacing on sediment accumulation

As sediment movement is interrupted, a beach builds both in plan and cross-section. When this strategy works, the beach builds up between groynes and longshore sediment transport is re-established. Sediment movement takes place either round the seaward end of the groyne or over the top of the groyne.

Groynes can be a very effective means of encouraging sediment to accumulate and so protect the coast. They represent a combination of hard (groynes) and soft (sand accumulation) engineering. However, beaches downdrift can be starved of sediment with the consequence that their plan and profile shrink. This can lead to accelerated erosion of the coast.

Soft engineering

The traditional reliance on engineering to provide solutions to issues of coastal erosion and flooding is being modified. This change in attitude should be seen in the context of:

- the increasing proportion of people living in coastal and estuarine locations
- the growing impacts of climate change, such as sea level rise and wave energy
- the understanding that many coastal zones experience sediment deficits
- the understanding that many coastal zones experience subsidence.

It is now recognised that instead of trying to resist the forces operating within the coastal zone, the approach more likely to succeed is one that seeks to 'work with the processes' to bring about more resilient and sustainable results.

Beach nourishment

Beach nourishment (also known as beach recharge) aims to improve the 'health' of a beach and so make it more resilient to out of the ordinary changes such as a severe storm. In addition, some schemes aim to improve the amenity value of a beach by making it more attractive to users.

Key elements of beach nourishment schemes are:

- a source of sediment nearby
- the impact of sediment removal by dredging/mining
- calibre of sediment equal to, or a little coarser than, existing sediment
- the techniques used to retain new sediment where it has been placed.

Costs vary depending on factors such as the method of bringing the sediment on to the beach and the distance this has to be transported. In general costs are between £5000 and £200 000 per 100 metres. Most large-scale nourishment schemes use sediment, typically sand, to be pumped onshore from dredgers (page 200).

A significant issue in assessing the success or otherwise of a beach nourishment scheme is the way that sediment is redistributed once it has been deposited. Wave energy can soon rework sediment, leading to the perception that the time, effort and money involved have not been worthwhile. However, beach systems tend to adjust relatively quickly through feedback mechanisms to an equilibrium.

 KEY TERM

Beach nourishment
Involves sediment being brought into a beach system to build up the plan and profile of the beach.

CONTEMPORARY CASE STUDY: THE DUTCH 'SAND ENGINE'

The Netherlands potentially faces considerable social, economic, environmental and political consequences of a relative rise in sea level. This is due to:

■ its densely populated area (500 persons per km²)

■ a long coastline – 350 km

■ subsidence of most of the coastal zone

■ over half its population, some 9 million, living below sea level

■ roughly 65 per cent of the country's Gross National Product (GNP) being generated within the coastal zone.

The Dutch have centuries of experience of managing their coastline and have developed almost unparalleled expertise in a variety of engineering approaches. The Sand Engine represents one of the latest innovative projects.

Traditionally, the Dutch have applied sand nourishment both to beaches and dunes as well as placing sand in the nearshore zone. The latter approach has been the more favoured one since the 1990s as it modifies processes such as wave breaking and sediment transport, resulting in a wider beach. Typically, such projects use about 1–2 million m³ of sand and have a lifetime of about 3–5 years.

The extensive damage caused by coastal flooding along parts of the USA's Gulf Coast, New Orleans in particular, following Hurricane Katrina in 2005, led to reappraisals of many coastal management plans. The Dutch recognised that they needed to increase substantially the annual volume of sand used for nourishment, from about 12 to as high as 80 million m³. Traditional methods of nourishment would require the entire length of the coastline to be engineered, resulting in beaches becoming far too wide from a recreational perspective.

The idea of a localised, mega-nourishment project, the Sand Engine, was born (Figure 7.12). It is located between The Hague and the Hook of Holland and was largely constructed between March and July 2011. Its key features are:

■ 21.5 million m³ of sand obtained from 10 km offshore

■ 2.4 km of coastline nourished

■ nourishment extends up to 1 km offshore

■ sand is deposited as a large hook-shaped spit with a base of 1 km attached to the shore

■ a 7.5 hectare lake

■ post-nourishment, sediment moves in a more natural fashion in both directions along the coast

■ ecological stress is minimised in a relatively small area where sand is initially deposited

■ the curved tip is designed to provide shelter from waves for the area behind

■ the shallow lagoon formed behind the sand spit provides habitats for marine organisms, e.g. flatfish

■ it is much cheaper (per m³) than repeated smaller scale nourishments, e.g. every 2 years.

Such large-scale projects are expected to have lives of between 10 and 20 years and to be more cost-effective in terms of construction costs versus benefits from coastal protection and no loss of recreational services. The project is still being assessed, such as by using jet ski-mounted surveying equipment, and so far sediment movements have resulted in increases in beach volumes so that some 200 hectares of beach have been gained. Responses from beach users, such as wave, wind and kite surfers, have been largely positive.

It will be a few more years before a comprehensive evaluation of the project becomes realistic. If it is successful, then more Sand Engines may appear along open sandy coastlines facing the risks from rising sea level.

▲ **Figure 7.12** Aerial photograph of the Sand Engine looking southwards. The image was taken in August 2016 and shows the location at low tide. The dispersal of sand northwards suggests that the scheme is functioning as intended. It is also interesting to note the recreational use being made of the sand by the number of people occupying the area closest to the sea

Crucial in the re-evaluation of management techniques is the recognition and increasing understanding of the role of sediment stores and pathways in the coastal zone. With the considerable advances in computing power (which are ongoing), various approaches to modelling sediment dynamics in ever more complex ways are actively being researched around the world.

Contrasting approaches to dune management

Coastal dune ecosystems are integral components of the coastal zone. When they are not healthy, risks to coasts are made worse.

Dune regeneration

Where natural dune systems are threatened, various methods are employed to try to restore a system's sustainability. Positive feedback processes often start with vegetation removal such as caused by a severe storm or intensive trampling. Fences, covering bare sand with brushwood or matting and replanting with marram grass have been used. At locations where marram grass needs an increase in fresh sand to grow, the deliberate creation of areas of bare sand for aeolian transport has been tried, such as in Braunton Burrows, North Devon.

Where pressure comes from human activities, restricting access, either completely or by providing clearly marked paths often with wooden planks, can allow natural processes to re-establish equilibrium.

Artificial dune construction

Along sandy coastlines where no natural dune systems exist, artificial dunes can be constructed. These are used on lowland coasts such as in the Netherlands and parts of the USA's East Coast. Structures designed to slow down wind speed are erected along the coast (Figure 7.13). Materials such as brushwood, fences made of thin wooden stakes wired together or woven fabric are used. The aim is to reduce airflow by about 50 per cent and to encourage sand accumulation either side of the fence. It is important that the fence is permeable to allow some air to pass through with its load of saltating sand. A solid fence would set up too much turbulence and reduce sand deposition. Once the sand has built up to almost cover the fence, either the fence can be lifted to a new height or another fence installed. Rates of accumulation of about 1 metre per year have been recorded. For this method to be sustainable, vegetation needs to be established in the system.

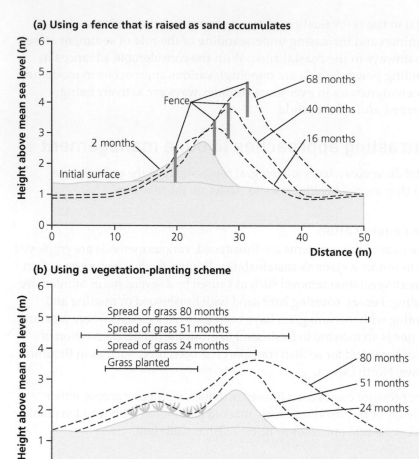

(a) Using a fence that is raised as sand accumulates

Height above mean sea level (m)

68 months
40 months
16 months
Fence
2 months
Initial surface

Distance (m)

(b) Using a vegetation-planting scheme

Height above mean sea level (m)

Spread of grass 80 months
Spread of grass 51 months
Spread of grass 24 months
Grass planted
80 months
51 months
24 months

Distance (m)

▲ **Figure 7.13** Artificial dune construction as used on barrier islands, North Carolina, USA

③ Planning for coastal change

▶ *How has improved knowledge and understanding of the coastal zone system influenced management of the coast?*

Protecting the coast from erosion and flooding has a long history with schemes that were local and piecemeal. There were no planning structures that could take an overview of a stretch of coastline. Certain events led to advances in technology, such as the invention of concretes capable of resisting marine conditions and the development during the Second World War of coastal engineering capabilities such as building temporary harbours. The impact on perceptions of risk of the catastrophic flooding around North Sea shores associated with the storm surge of 1953 should also not be underestimated. However, there were no strategic approaches with planning and management largely focused on a single issue that was dealt with by a single authority. A coastal town that had undergone some flooding, for example, would have its sea wall raised by the town council. Elsewhere, the reclamation of salt marsh for agriculture or development was the priority.

Changing ways coastal management decisions are taken

Even if the original issue for a particular intervention (e.g. sea wall or groynes) was resolved, unintended consequences frequently arose. The growing awareness of interactions among the natural components of the coastal zone, as well as between the natural and human, has led to significant developments in the ways coasts are planned and managed.

The increased complexity of issues arising in the use of coastal locations and the acknowledgement of the various scales, spatial and temporal, that coastal management must take into account, has resulted in new structures for decision-making (Figure 7.14).

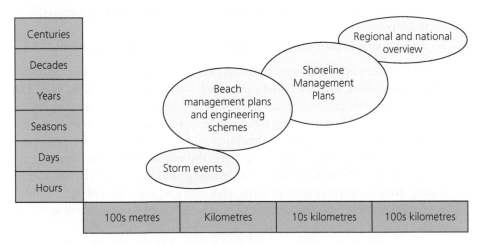

◀ **Figure 7.14** Coastal management at different spatial and temporal scales

If the intention is to manage a shoreline, then the appropriate spatial and time (temporal) scales over which changes in the shoreline happen need to be considered. While different, the types of management overlap, which reflects the operation of natural processes and landform development.

Integrated Coastal Zone Management (ICZM)

With the increasing attention being given to coastal matters towards the end of the twentieth century, the idea emerged of a different approach to coastal management. Previously, issues such as cliff erosion, flooding and the health of ecosystems was considered in a piecemeal way, with analysis and decisions taken at local scales, such as individual towns. The 1992

Earth Summit held in Rio de Janeiro encouraged all types of environmental management to have a holistic approach. For coasts this meant considering the multiple interactions that might operate, both natural and human.

As well as acknowledging the importance of understanding flows of energy and materials within and among coastal zone systems, ICZM incorporates the idea of sustainability. It is important to appreciate that this was not simply aimed at coming to a set of 'sustainable' actions but rather that it refers to a way of thinking or a state of mind. It encourages issues to be considered not just in the present but for the future. In geographical terms it also means considering matters of spatial scale, understanding that there is a hierarchy of processes and landforms, such as:

Berm → Beach profile → Entire beach → Bay → Headland → Bay complex → Coastal zone

Integration is meant to apply in several ways:

- among all elements in the physical environment
- among sectors of human activity, e.g. fisheries, tourism, settlement, transport
- among levels of government – local, regional and national
- among countries – sediment transport does not stop at international borders
- among perspectives such as scientific, cultural, political, economic.

ICZM has been adopted by many countries, including those in the EU, New Zealand and Iran. While a holistic approach has many advantages, as ICZM has been put into practice, it has tended to be a 'top-down' approach. This is because it has been operated by institutions with administrative authorities having the legal responsibility and controlling the money to pay for schemes. A significant aspect of this is the question of ownership of resources such as the water within an estuary, sediment stores in the coastal zone or communities of birds, shellfish and fish. As with all management schemes, the key test is how well it functions in the real world.

Shoreline Management Plans (SMPs)

Towards the end of the twentieth century, integrated plans for coastline management in England and Wales were given an official structure based largely on the sediment cell systems (page 44). Various organisations (for example Environment Agency, local councils, Department for Food and Rural Affairs, Natural England, National Trust) were involved in gathering information, consulting, planning and agreeing a policy for the length of shoreline they are responsible for. Various reviews of the SMP system have been carried out to refine how they operate. The primary aim is to identify the most sustainable approaches to managing the flood and coastal erosion risks to the coastline across three time scales:

- Short term – 0 to 20 years
- Medium term – 20 to 50 years
- Long term – 50 to 100 years.

A strong signal in the reviews has been that SMPs must be 'realistic' and not promise what cannot be delivered in economic and technical terms. One issue is that even if a local authority adopts an SMP, this does not represent a firm commitment to fund the implementation of those policies. The importance of extending the assessment of risks to the long term is that SMP planning must be flexible enough to adapt to changes in areas such as legislation, politics and social attitudes, such as perceptions of flood risk among local communities and attitudes towards funding coastal defence. It is important that the long-term vision for an SMP is not adversely affected by decisions made in the short term, for example avoiding committing future generations to inflexible and expensive defence.

What is sustainability?

In general terms, if something is 'sustainable' it is capable of being supported or upheld. Sustainability has become a theme applied to all sorts of areas of human activities, becoming something that gives automatic credit to anything it is applied to. However, such unquestioning acceptance can act as a smokescreen that obscures some fundamental issues.

The definition of sustainable development as '… development that meets the needs of the present without compromising the ability of future generations to meet their own needs' became popular after the 1987 World Commission on Environment and Development (WCED). It is often referred to as the Brundtland Report after its Norwegian chair, Gro Harlem Brundtland. However, there are numerous interpretations of this apparently straightforward definition. It is important to understand that sustainability is a contested concept, that is there are very different views of what it should mean.

The Brundtland Commission has been heavily criticised for its reliance on technology, its inadequate attempt to address population growth and the absence of a focus on the unsustainable practices of transnational corporations (TNCs). Critics put forward the claim that the aim was to sustain development rather than to sustain ecosystems or natural processes. Organisations such as the World Bank which embraced Brundtland were included in the criticism.

The 1992 follow-up to Brundtland held in Rio de Janeiro also placed sustainability in the context of economic growth. Different views of the achievements of Rio exist but perhaps its importance can be seen in the continuing, some would say expanding, roles that environmental issues have.

It is important that terms such as 'sustainable' should not simply be taken at face value but should be explored and thought about from different perspectives. What is sustainable in one place at one time may not be in another location at the same or different time.

Within the 22 larger-scale SMPs, smaller-scale management units have been identified so that policies appropriate for localised locations can be identified and put into action. Four such policies are considered, known as Strategic Coastal Defence Options (SCDOs) (Table 7.1).

SCDO option	Actions
1 Advance the existing defence line	Construct additional defences seaward of the current coastline; most likely to be achieved by land reclamation
2 Hold the existing defence line	Maintain and/or strengthen existing defences to prevent any flooding or erosion; rebuild sea walls, raise the height of flood defences, e.g. embankments; install floodgates; beach nourishment; maintain groynes
3 Managed realignment	Let natural processes operate to allow the development of landforms such as salt marsh which can then act as defences; usually involves some loss of formerly protected land but in a controlled way, such as creating gaps in embankments to let seawater flow into former grazing marsh
4 No active intervention	Natural processes are not prevented from operating so that the coastal zone system functions with no interference; can be applied to areas currently defended that will no longer be protected, as well as areas with no defences

▲ **Table 7.1** Coastal defence options

ANALYSIS AND INTERPRETATION

Study Table 7.2, which shows the extent of retreat of the coastline at Walton-on-the-Naze, Essex, eastern England, 1300 to 2000.

Time period	Land lost (km)	Rate of loss (metres/year)
1300–1600	4.8	
1601–1777	0.8	
1778–1950	0.6	
1951–2000	0.1	2

▲ **Table 7.2** Rates of coastal retreat at Walton-on-the-Naze, Essex, eastern England, 1300–2000

(a) Using Table 7.2, calculate the rate of loss of land in metres per year for each of the three missing time periods.

GUIDANCE

Although straightforward, there are some points to be careful with in this calculation. The land lost is given in kilometres so must be converted to metres. Then the number of years in the time period needs to be calculated so that the rate of loss in metres per year can be calculated. As a generic point, it is always important to be aware of the units that data are given in and to be clear as to the units any answer is to be given in.

(b) Using Table 7.2, suggest reasons for variations in the rate of loss of land since 1300.

GUIDANCE

It is helpful first to state in broad terms what the pattern of loss is. This provides the framework on which to build a response dealing with possible reasons. Over the time from the fourteenth century through to the present day, the rate of loss of land has decreased. Considering the factors influencing the cliff system (page 54) will help keep the response focused on cliff retreat. Clearly this stretch of coastline has eroded at a relatively rapid rate, with even the rate of loss of the past few decades high compared to many locations. It is therefore unlikely that measures aimed at slowing or preventing erosion have been implemented, such as cliff drainage or rip-rap at the cliff base. As cliffs are eroded by a combination of marine and sub-aerial processes, the former cliff face falls to the base of the cliff. Here the material is further broken down until it is of a sufficient calibre to be transported by wave action. As cliff retreat proceeds, a shore platform develops. Over time, the level of wave energy able to reach the cliff face diminishes, thereby reducing the effectiveness of marine erosional processes such as abrasion. This leaves sub-aerial processes as the dominant force acting on the cliff. It is this reduction in wave energy (due to the increasing width of the shore platform) into this particular cliff system that may be mainly responsible for the slowdown in rate of retreat.

(c) Examine the value of a sustainable management approach to coastal protection.

GUIDANCE

The essence of 'examine' is to consider carefully, in this case the value of sustainable management for protection of the coast. Firstly, some discussion could be given to what is meant by 'sustainable'. This is not a straightforward concept as its meaning is contested, that is there are different and sometimes opposing views as to what it should include. For example, it might be that a stretch of vulnerable coastline has housing close to the cliff top. In the short term, the next decade, it may be desirable to protect the cliffs from further collapse. However, over the longer term, the next 50 to 100 years, maybe the costs of defending the cliffs cannot be justified. It is also worth considering different areas of sustainability – economic, social and environmental. A coastal protection scheme may be economically sustainable but the impacts lead to environmental degradation such as the disruption of sediment movement within a cell. Thinking about 'value' can pick up this theme of different types of sustainability. Is it possible to compare the value of protecting homes or infrastructure such as transport routes with the value of the aesthetics of a landscape? What is lost and gained when coastal protection involves hard, soft or no engineering?

CONTEMPORARY CASE STUDY: SHORELINE MANAGEMENT IN ACTION, EAST DEVON

This stretch of coastline is part of the South Devon and Dorset SMP which is subdivided into small-scale policy units, of which there are 28 in and around the Exe estuary. In this area, Policy Scenario Area 8 (PSA8), there are several large, such as Exeter, Exmouth and Dawlish, and many smaller, settlements. The Exe estuary dominates PSA8 and is a crucial area environmentally. It is a Special Protection Area (SPA), a **Ramsar** site and an **SSSI**, being an internationally important site for bird populations which find food and shelter in the wide variety of estuary habitats. Both estuary banks and part of the coast to the south-west have railway lines: one an important commuter route from Exmouth to Exeter, the other the main rail link between the south-west region and the rest of the national rail network.

The designation of a location as a Ramsar site is an example of how successful global governance can be. It was the first treaty between nations aimed at conserving natural resources, brought into being in 1971. Securing international agreements and enforcement is necessary as some locations spread across national boundaries, such as lakes and deltas, and the migration paths of many species, such as wildfowl, cross national borders.

The policies adopted by the latest revision of the SMP for the estuary and adjacent coastline are overwhelmingly 'Hold the existing defence line' along with some sections where 'No active intervention' applies. These apply over all three time scales, short,

medium and long term, although the possibility of managed realignment has been raised as something to consider.

Significant issues within PSA8 are:

- how to manage the spit at Dawlish Warren
- coastal squeeze along those stretches where defences are to be maintained
- possible managed realignment at locations to create new habitats to compensate for where coastal squeeze has caused a loss
- long-term sustainable defence of open coastline at Dawlish and Exmouth.

 KEY TERMS

Ramsar The International Convention on Wetlands of International Importance is known as the Ramsar Convention. It is an international treaty (and an example of global governance) providing a framework for national and international action and cooperation for the conservation and use of wetlands.

SSSI A Site of Special Scientific Interest is a location designated due to its outstanding scientific importance. SSSIs can be geological or ecological in importance.

In order to understand something of the complex interactions which need to be taken into account when managing stretches of coastline, it is important to appreciate their physical background.

CONTEMPORARY CASE STUDY: THE EXE ESTUARY

A local-scale, low energy coast, the Exe estuary basin is a long-established feature of the area, possibly dating back to the late Tertiary period (c. 2–5 million years ago). With the fall in sea level during Pleistocene glaciations, river channels excavated deeply into the estuary and extended out to what is now the sea. These were filled with fluvial sediments as conditions warmed at the end of the last ice age.

As sea level rose during the Holocene period (page 117) vast quantities of sediment were moved onshore. This sediment was the result of fluvial and periglacial processes operating over the extensive area exposed due to the fall in sea level. The rise in sea level slowed and about 6000 years ago a coastline very similar to the one evident today came into existence.

▲ **Figure 7.15** Mouth of the Exe estuary looking to the north-west, with the distal end of Dawlish Warren on the left and the town of Exmouth on the right. The upstream direction of the Exe is to the right

Today, the coastal zone including the estuary is a complex system of landforms and human-made features:

- Exe estuary – a 3000-hectare ria, some 13 kilometres long.

- Dawlish Warren spit – a 2-kilometre sand landform extending across the western side of the estuary mouth formed by shoreward and westerly longshore transport during the Holocene; limited supply of fresh sediment from Holcombe cliffs and Dawlish Beach to the south-west due to the sea wall and railway line running along the shore forming a barrier to marine erosion and transport.

- Exmouth spit – a small sand landform (800 metres) extending eastwards across the eastern side of the estuary mouth in a similar way to the Dawlish Warren landform. Energy for sand transport came from easterly winds and waves. Now built over by part of the town of Exmouth the shoreline is completely defended by a substantial sea wall and promenade.

- Exmouth beach – a 3-kilometre sandy beach extending eastwards from Exmouth docks at the estuary mouth. A small sand dune system was destroyed by severe storms in the 2013/14 winter. There is some supply of sand from Orcombe cliffs to the east.

- Sand banks – Pole Sands lie just seaward of the estuary mouth, Bull Hill Bank lies just landward of the estuary mouth. These two sediment stores have a dynamic movement of sediment in and out depending on tide and river currents and wave energy.

- Mud and sand flats and salt marsh within the low wave energy environment of the estuary. These are sediment sinks for material brought down by the rivers and carried in by the flood tides (regular) and storm surges (infrequent).

Sediment movements are key interactions among landforms within a low energy environment (Figure 7.16).

There is a flow of mostly sand along the coast from the south-west and east and a wave-driven flow of sand from

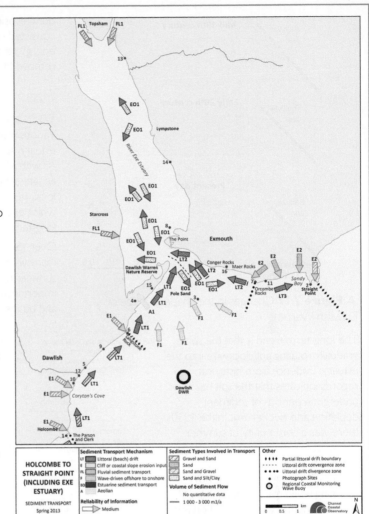

▲ **Figure 7.16** Sediment movements in and around the Exe estuary

the offshore sand banks at the mouth of the estuary. Additionally there are inputs of sediment from fluvial sources, some brought down by the River Exe and the rest transported by smaller tributaries that enter the estuary along its sides. Sediment also moves within and out of the estuary. Waves, tides, river discharge and sediment, together with human interventions, interact to give the characteristic landforms and overall landscape of the area.

Dawlish Warren spit is the most significant landform within the landscape of the Exe estuary. Its geomorphological history, as charted by contemporary maps and written accounts in the past and modern surveying techniques today, reveal a dynamic landform (Figure 7.17).

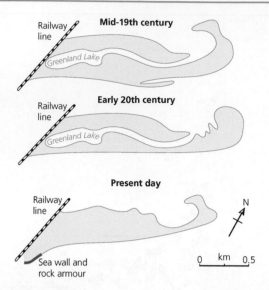

▲ **Figure 7.17** Changes in the shape and size of Dawlish Warren

The long-term trend is that the spit is gradually rotating anticlockwise into the estuary. Evidence from historical records indicates that the spit has undergone periods of sediment depletion and **accretion**. In the 1930s there was a community of summer chalets at the distal end but by 1946 these had been washed away by storms. The eastern 'hook' is now accreting while the southern shore is eroding, and the very thin section is in real danger of being overtopped by a storm event. The proximal end is defended by substantial sea walls and rock armour.

This stretch of coastline does not receive the high average energy inputs that some more rocky areas do, and deposition is a dominant process in the estuary, such as the accumulation of tidal flats and salt marsh. One should not, however, underestimate the energy flows in terms of the ecosystems in the estuary. It is also the case, that considerable movements of sediments occur and that the occasional high energy event can result in considerable geomorphological 'work' in the form of modifying landforms such as

dune erosion. Part of the coastline at Exmouth once consisted of some sand dunes but several high energy storms in the past 15 years have largely removed them.

Present-day management of the Exe estuary

Various plans and policies are operational within the estuary, some are complementary while others conflict. As with all such locations, a variety of stakeholders have an interest in the estuary. These stakeholders have local, regional, national and international perspectives which exist across several time scales, short- to long-term (Figure 7.18).

Over 150 000 people live close to the Exe estuary with many visitors swelling this number, especially during the summer holiday season. The estuary is of international importance because of its wildlife. Many communities and businesses have established along the coastline and

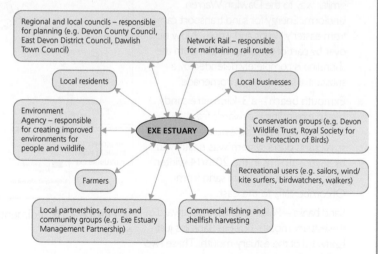

▲ **Figure 7.18** Stakeholders in the management of the Exe estuary

together with farmers and transport routes on both west and east banks of the estuary, rely on defences against erosion and flooding. These defences have been reviewed and are reflected in the current SMP policies (pages 195–6).

 **KEY TERM**

Accretion The gradual building up of something.

The Dawlish Warren beach management scheme – a combination of hard and soft engineering

The greatest challenges concerning the management of Dawlish Warren are:

- The seaward-facing sand dunes are eroding significantly along most of the spit's length.
- The Special Area of Conservation is being damaged by the gabions installed in the 1960s and 1970s.
- If the narrow neck were to be broken through, the increase in wave energy and water depth in the estuary would threaten humans (housing, businesses, transport, agricultural land) and wildlife (habitats).

The strategy, established in May 2014, focuses on actions required by 2030 but with an overview looking ahead for the next 100 years. Hold the Line is the approach for the area around the settlement of Dawlish Warren and along the railway route. The strategy also suggested that the spit should continue to act as a barrier to storm waves. In addition the beach volume could be raised and this would in turn allow the dune system to recover as less wave energy would reach the dune front (Figure 7.19).

The key features of the scheme include:

- Cost of £14 million → estimated benefits of £158 million
- Started January 2017 – completed October 2017
- 250 000 m³ sand pumped on to beach from Pole Sands
- 460 metre-long Geotube buried in dunes across the spit neck; it consists of sand pumped into giant geotextile (woven polyester that is permeable) bags 2.85 metres high
- Timber groynes lengthened and strengthened (Figure 7.10)
- Fencing installed over dunes to aid their recovery.

It was always known that as soon as the scheme was complete, energy flows through the coastal zone would start to bring about feedback. Sand is being moved, beach profiles and plan are altering and the dunes are adjusting. The medium- and long-term SMP for the spit is to allow for Managed Realignment and No Active Intervention, although requirements may change over a time scale as long as this. The effects of climate change over the medium and longer term are not going to be straightforward to manage.

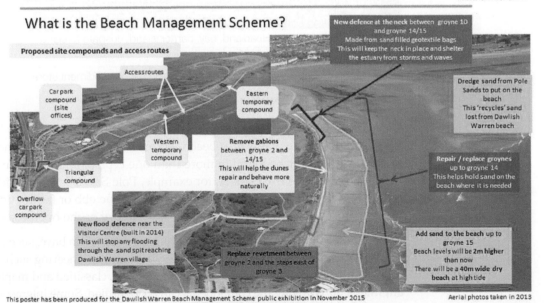

▲ **Figure 7.19** The Dawlish Warren Beach Management Scheme

A new approach to modelling coastal evolution – CESM

To date, coastal management has relied upon the concept of the sediment budget in the context of hierarchies of littoral or sediment cells (page 44). The use of these natural coastline units represented a considerable improvement over a reliance on administrative boundaries such as counties. However, limitations of the sediment cell approach are now being recognised:

- cells mainly reflect short-range transfers of medium- and larger-sized sediment
- they do not reflect movements of finer-grained sediment, especially over longer distances
- there are concerns over how the cells are delimited
- there are concerns as to how stable cell boundaries will be in the future as sea level rises and wave energies alter
- there are concerns about changes in sediment supply into and within the coastal zone.

One approach beginning to be explored is that of Coastal and Estuarine System Mapping (CESM). This brings together open coastlines, the inner offshore shelf and estuaries that have previously been treated as separate units. A key driver for this change has been the need to consider how coastal and estuary systems interact and change at scales that are crucial for management.

Landform complexes, groups of landforms, are identified within each of the three primary coastal units.

Primary coastal environment	Examples of landform complexes
Open coast	Headland, bay, barrier island, cuspate foreland
Estuary	Fjord, ria, tidal inlet
Inner shelf	Offshore sandbanks and various sediment stores

▲ **Table 7.3** Primary coastal units and their landforms

Each landform complex possesses a set of landforms. For example, a bay may contain a shore platform, a beach, berms (ridges) and dunes. The same landform may be found in more than one type of landform complex. At the mouth of the Exe estuary, for example, Pole Sands is an ebb tidal delta (an accumulation of sediment deposited by the ebb or falling tide). This landform can be considered part of the estuary or the open coast.

CESM also recognises the important role human actions can have, such as coastal defence structures (e.g. groynes, sea walls) and engineering such as beach nourishment. Once these have been identified, classified and mapped, any interactions among the components can be identified. Some interactions involve modifying a process, for example a breakwater shelters an area from wave action, while others involve sediment flows, cliffs and beach.

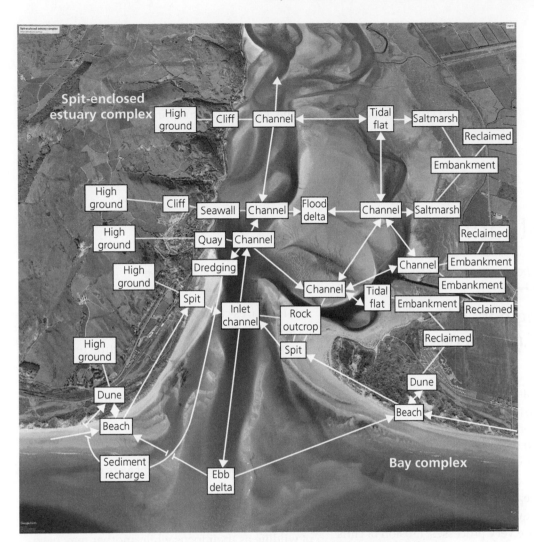

▲ **Figure 7.20** CESM applied to a spit-enclosed estuary, highlighting the landforms and interactions present. Human interventions in the coastal zone are in black text boxes

Once the various elements have been identified they can be put into a Geographic Information System (GIS) which allows layers of data, for example geology and water depth, to be added or subtracted as the analysis proceeds. Another key aspect of the CESM approach is to construct matrices that portray the probability of interactions between elements, such as sediment flows between a beach and an estuary. CESM also allows the identification of different future scenarios for a coastline, such as the breaching of a barrier and the possible effects on different landforms.

 KEY TERM

Geographic Information System (GIS) An integrated computer tool for storing, processing and analysing spatial data.

Potentially the most important aspect of the CESM approach is that it allows non-scientists a role. Traditionally, coastal management has been a 'top-down' approach from consultants, engineers and planners, even when public consultations have been held. Because the software developed for CESM is open source, available to anyone, communities can come together to create a model of a coastline. Because the coastline faces decades of progressive change, the more that people living along the coast are involved in the planning and management of the coast, the greater the likelihood of appropriate decision-making.

④ Evaluating the issue

▶ *Discussing the view that hard coastal defences cause more problems than they solve.*

Identifying possible contexts for the study of coastal defences

With the increasing intensity of human activities within the coastal zone, in particular the accumulation of settlement, transport routes, agriculture and recreation along coastlines, interactions between the natural environment and people can cause issues. One of the issues is how to manage the physical processes operating along the coast when human activities are put at risk due to those very processes. Erosion of cliffs and flooding by seawater are concerns facing coastal communities around the world.

Traditionally the approach to defending human activities from marine processes, such as wave energy and storm surge, has been to construct substantial 'hard' structures, designed to keep the sea at bay. These have allowed a wide variety of human activities to exist and flourish, but in recent years a reappraisal of how best to interact with the sea has been taking place.

Another theme for consideration is the way that the 'hard' management of particularly important coastal places can involve the intersection of local-, national- and even global-scale governance issues, potentially creating tension or conflict. The discussion should also consider the basis on which some places are defended while others are not. In the Shoreline Management Plans for the England and Wales coast, the designation of which of the four approaches is adopted for any one segment is complex. The involvement of a diversity of players with contrasting perspectives on the way to manage the coast can make the decision-making process contested. One player may have the conservation of wildlife as their focus while another may have agricultural interests to promote. And there is always the issue of where the money for management is to come from.

Deciding on the 'value' to allocate to what is to be protected or not can be viewed in different ways. Should defending economic activities be considered of greater importance than protecting the landscape of a location? How much weight should be given to the conservation of ecosystems? It is in this context that the increasing attention and monetary value given to ecosystem services is of considerable significance.

Added into the mix for some locations are issues concerning international governance. The role of an international organisation such as UNESCO in determining World Heritage Site status, for the Jurassic Coast for example, adds a complicating factor to the local- or regional-scale concerns of coastal defence.

View 1: Hard defences help solve problems arising from coastal systems

The construction of hard structures has given opportunities for various human activities that have benefited people. To begin with, simple wooden quays allowed boats to tie up and earth embankments led to grazing meadows for cattle and sheep. As centuries passed, and in particular from the late eighteenth century on, rising engineering ambition and capabilities allowed large-scale schemes to be undertaken. Port cities such as Liverpool, Hamburg, Singapore and Los Angeles have grown as a result of hard engineering along the coast. The same principles of defence against the sea have led to settlements right across the urban hierarchy existing and providing homes and places to work for millions of people. Places experiencing great pressure for space, such as Hong Kong or Singapore, rely on hard defences to allow land reclamation to proceed and thereby give people homes to live in and places for work. People living inland also benefit from sea walls, docks and jetties. They can trade through a port, importing and exporting goods. The process of globalisation depends on sea ports with their systems of hard engineering and associated infrastructure.

Tourism and recreation have benefited from sea walls and promenades, allowing resorts to grow, very often on land that previously was marshy or unstable. Beaches have been stabilised by groyne construction, thereby accumulating sand to encourage visitors.

Flood risks have always been significant along coastlines, especially low-lying ones such as the eastern coasts of the USA and England and the Low Countries (the Netherlands and Belgium). The many kilometres of hard defences protect millions of people and huge numbers of farms, factories and other types of economic activities. Low-lying island states are turning to hard engineering to offer them at least some medium-term security against rising sea level. Island states in both the Pacific and Indian Oceans, such as the Maldives, are actively constructing hard defences to mitigate against rising sea level.

View 2: Hard coastal defences can create new problems within a coastal systems framework

When first installed, hard defences such as sea walls seemed to have given people the upper hand in a battle against the sea. However, as the decades passed, it became apparent that the relentless action of waves and spray eventually wore away the concrete, stone and steel so that defences were breached. In places, so significant was the investment in the built environment protected by hard engineering that communities were left with no choice but to repair and reconstruct defences. A classic example is the railway line along the Devon coast past Dawlish. Built by the engineer genius Brunel in the mid-nineteenth century, it is now increasingly vulnerable to marine forces when storm events occur. There is no alternative at present but to commit resources to maintaining the track as this the only rail route into the West Country beyond Exeter.

Many groynes were constructed in the latter quarter of the nineteenth century and throughout the twentieth century in countries such as the UK, France, USA and Japan. Their success at slowing the rate of longshore drift and trapping sediment was clear and served exactly the purpose for which they were installed. As we now appreciate, but was not understood until relatively recently, locations downstream of the sediment flow were left exposed to wave attack as their beaches shrank in width and cross-section. Cliff erosion at Barton-on-Sea and the threat of overtopping at Hurst Castle spit a little further east can be linked with the installation of a substantial groyne field along the Bournemouth beaches. Even a single groyne can impact an area in ways that were not the intention of those who built it in the first place.

Many sea walls have been very successful, as has already been noted. However, there is always the issue of what happens where the wall ends, as wave energy can cause erosion at and behind the end section of the wall.

Hard coastal defences use materials such as concrete, steel and granite boulders. Concrete and steel manufacturing releases vast amounts of greenhouse gases, contributing to rising sea surface temperatures and the thermal expansion of seawater and thereby raising sea level. Granite boulders can be transported considerable distances from where they are quarried to where they are deployed as rip-rap defence. In the UK many coastal defence schemes use Norwegian stone, therefore incurring considerable carbon miles.

View 3: Hard defences can create unexpected governance problems when managing coastal places

For some stretches of coast, international players are involved. Places such as the Great Barrier Reef and the Jurassic Coast are UNESCO-nominated World Heritage Sites – a highly sought-after, and then jealously guarded, status. As a contribution

to placemaking and potential rebranding processes, such international recognition is just about second-to-none. However, these designations come about because of the natural processes, landforms and landscapes located in the site. There is, therefore, a pre-eminent requirement not to allow management that might threaten this status, which means that approaches to coastal defence are not straightforward. At places such as Lyme Regis and Charmouth on the Jurassic Coast, a delicate balance is aimed at between defending the built environment of the town from coastal erosion and flooding while allowing the cliffs close by to be eroded. The placemaking process is reliant largely on the continuing supply of fossils for which the area has a worldwide reputation.

Arriving at an evidenced conclusion

There is a temptation today not to recognise the substantial advantages and benefits that hard coastal defences have given to millions of people. It is important to point these out and to recognise that when many hard engineering schemes were planned and installed, the processes operating within the coastal zone were little understood. This does not take anything away from the point that some damaging unintentional consequences have arisen.

There is also the tendency today to use sustainability as a justification for condemning or promoting certain courses of action, but without considering what is actually meant by something being 'sustainable'. Is it economically sustainable to remove a groyne field, allow sediment to move unhindered along a coast and so lose a resort's beach? Is it socially sustainable not to protect low-lying coastal areas where many thousands of people live and work and so force them to move away? To what extent is it important to act in ways that do not put at risk a designation as significant as the Jurassic Coast's World Heritage Site status or the Exe estuary's Ramsar status?

There are no easy solutions in coastal management. Rising sea level is a reality in the short, medium and long term. The many millions of people living along coastlines are a reality.

There is an argument to be made about the environmental damage some hard coastal defences have been responsible for. But perhaps the key conclusion is that we need to acquire more substantial knowledge and authoritative understanding about coastal patterns and processes in order to manage coastal areas more successfully.

Chapter summary

✔ Coastlines are hazardous environments for human activities where several factors interact to bring about risks from erosion and flooding; a growing issue concerns rising sea levels as a consequence of global warming. The erosion of cliffs can threaten buildings and infrastructure such as transport routes.

✔ Some geologies represent particularly high risks, such as clays, some sandstones and recently deposited materials such as from glacial activity. However, assessing the rate at which cliffs are receding is not straightforward. Historical records can be inaccurate and unreliable and average rates hide sudden high energy events.

✔ Hard engineering has been deployed to resist wave energy and high tides that threaten to flood coastal locations. Marine cliffs are also protected from sub-aerial weathering and mass movements using various techniques. Sea wall design has evolved to offer more appropriate methods of protection; some designs aim to absorb and dissipate wave energy. However, where a hard defence line is constructed, for example a sea wall or embankment, coastal squeeze can become a concern and lead to loss of habitats, for example salt marsh

✔ Sediment accumulation, for example through the use of groynes, is designed to absorb and dissipate wave energy. Various soft engineering techniques are applied, very often intended to promote sediment build-up, for example beach nourishment and sand dune conservation.

✔ Coastal management structures are evolving to fit better with a deeper appreciation of the time and spatial scales coastal processes operate over. More holistic approaches are now common so that sediment inputs, stores flows and outputs are receiving much attention in order to put into action more effective coastal management schemes. Increasingly, the relations among the landforms making up the coast, including estuaries, are being recognised as vital to successful coastal management.

Refresher questions

1 Outline reasons why coasts are risky environments for people.

2 Explain why it is not straightforward to assess rates of cliff recession.

3 Explain how cliff recession is a result of the interaction of several factors.

4 Describe and explain the difference between 'hard' and 'soft' engineering as applied to coastal management techniques.

5 Outline the advantages and disadvantages of sea walls in coastal defence.

6 Explain how 'coastal squeeze' can arise and why it poses a risk to sustainable coastal management.

7 Explain why defending coastal cliffs from sub-aerial weathering and mass movements might be a disadvantage to coastal defence.

8 Describe the advantages and disadvantages of beach nourishment.

9 Outline the advantages of Integrated Coastal Zone Management.

10 Suggest the advantages of the Coastal and Estuarine System Mapping approach to coastal management.

Discussion activities

1 In small groups, obtain the details of the Shoreline Management Plans either for your local coastline or for one that you are studying and perhaps have visited on a field course. Use Ordnance Survey maps (1:50 000 and 1:25 000 scales) and your own knowledge and understanding of coastal processes to describe the principal landforms and processes operating along the coast. Suggest reasons for the actions proposed in the SMPs, bearing in mind the range of human activities shown on the OS maps. Consider the implications of not defending any areas that currently are being protected.

2 Divide your class into two. Everyone in one half should prepare a speech advocating the advantages of hard engineering in coastal management, while the other half prepare speeches promoting soft engineering. In opposing pairs, present your respective arguments. Each person should then refine their arguments and then a whole class debate should be held. At its conclusion, each person should write an analysis of the pros and cons of both types of approach to coastal management. Remember to consider economic, social and environmental factors.

3 Research two or three coastal management schemes from around the world, choosing from

countries at different points along the development continuum. Evaluate their respective advantages and disadvantages, making sure that you take into account the contrasting development context the schemes are set in.

4 For a local area undergoing erosion and retreat of the coastline, consult several map sources going back in time as far as you can. Old large-scale Ordnance Survey maps (1:10 000, 1:2500, 6-inch) show the coastline and defences in detail. Historic aerial photographs and postcards, local newspaper reports and some academic sites such as coastal universities, also can provide valuable information on coastline erosion and retreat. On a sketch map construct the positions of former coastlines and discuss the factors that led to the pattern of erosion. Remember to consider the reliability and accuracy of the sources you consult and acknowledge these in your discussions.

5 In small groups, choose a different coastal location for each group. It needs to be an intensively used stretch of coastline such as southern California, the area around the Rhine delta or the Sydney Bay region for example. For your selected area, discuss the factors that have brought about the high intensity of human activity. Suggest the key issues facing the management of the coastline today and for the next 50 years or so, taking into account the predicted rise in sea level.

FIELDWORK FOCUS

A *Investigating the effectiveness of a management scheme using groynes could be the basis of an A-level investigation.* Measuring beach profiles (page 75) and beach accumulation on either side of a set of groynes would provide useful data in your assessment.

B *Comparing and contrasting cliff profiles at different sites along managed and unmanaged stretches of coastline can give indications of the impacts of coastal management schemes.* Photograph analysis from the present day as well as historical evidence such as old postcards (frequently found in second-hand bookshops or miscellaneous second-hand retailers) as well as formal publications such as town histories can offer valuable material. Estimating cliff heights using basic trigonometry can be helpful.

C *The impacts of a management scheme might be investigated in terms of gathering the views and opinions of various stakeholders.* Not only might views be sought from resident stakeholders but these could be contrasted with those of visitors, for example during the summer holiday season. Such investigations could also be linked with an investigation into informal representations of place, focusing on the ways a coastal management scheme might impact perceptions of different groups. Both qualitative and quantitative data could be valuable.

Further reading

Dawson, D., Shaw, J., Gehrels, W.R. (2016) 'Sea-level rise impacts on transport infrastructure: the notorious case of the coastal railway line at Dawlish, England', *Journal of Transport Geography*, 51, pp.97–109

Environment Agency – Shoreline Management Plans [Open access SMPs available online at www.gov.uk/government/publications/shoreline-management-plans-smps/shoreline-management-plans-smps]

French, J.R., Burningham, H., Thornhill, G., Whitehouse, R., Nicholls, R.J. (2016) 'Conceptualising and mapping coupled estuary, coast and inner shelf sediment systems', *Geomorphology*, 256, pp.17–35.

Harvey, B., Caton, N. (2010) *Coastal Management in Australia*. Adelaide: University of Adelaide Press. [Free e-book: www.adelaide.edu.au/press/titles/coastal/Coastal-eBook.pdf]

Hill, K. (2015) 'Coastal infrastructure: a typology for the next century of adaptation to sea-level rise', *Frontiers in Ecology and the Environment*, 13(9), pp.468–76

Nicholls, R.J., Townend, I.H., Bradbury, A.P., Ramsbottom, D., Day S.A. (2013) 'Planning for long-term coastal change: experiences from England and Wales', *Ocean Engineering*, 71, pp.3–16.

Stive, M.J.F., de Schipper, M.A., Luijendijk, A.P., Aarninkhof, S.G.J., van Gelder-Maas, C., de Vries, J., de Vries, S., Henriquez, M., Marx, S., Ranasinghe, R. (2013) 'A New Alternative to Saving Our Beaches from Sea-Level Rise: The Sand Engine.' *Journal of Coastal Research*, 29(5), pp.1001–8

CHAPTER 8

Study guides

① AQA: Coastal Systems and Landscapes

Content guidance

The optional topic of Coastal Systems and Landscapes is supported fully by this book.

The Coastal Systems and Landscapes option focuses on coastal zones. To help structure the study of dynamic coastal environments a systems approach is specified by AQA and is used throughout this book. This approach allows students to identify and investigate how fundamental geomorphological processes operate and interact among themselves and with materials such as sediments in the coastal zone. The outcomes are distinctive landforms and landscapes. The relationships between process, time, landforms and landscapes are to be explored in coastal settings.

The importance of the coastal zone to human activities and habitats can be appreciated as students investigate the various features located along coastlines. Study of the coastal zone also offers opportunities to exercise and develop a variety of skills, such as observational, measurement, mapping, data manipulation including presentational and statistical skills. Fieldwork in the coastal zone offers many opportunities for students to practise both their quantitative and qualitative skills.

Coastal Systems and Landscapes (Topic 3.1.3)

Sub-theme and content	Using this book
3.1.3.1 Coasts as natural systems The topic begins by considering systems concepts, looking at aspects such as inputs, stores, flows/transfers and outputs. Concepts such as feedback (positive/negative) and equilibrium are important. Students also consider how related landforms combine to form characteristic landscapes.	Chapter 1 pages 2–22
3.1.3.2 Systems and processes This section focuses on the components of the coastal system. ■ Energy sources in coastal environments: winds, waves, currents and tides ■ Sources of sediments, sediment cells and budgets	Chapter 1 pages 5–18 Chapter 2 pages 36–9, 43–5
The fundamental geomorphological processes of weathering, mass movement, run-off, erosion, transport and deposition are to be understood in general as well as being given an explicitly coastal context: ■ Erosion – hydraulic action, wave quarrying, corrasion/abrasion, attrition	Chapter 2 pages 29–36
■ Weathering – cavitation, solution ■ Transportation – traction, suspension (longshore/littoral drift)	Chapter 2 pages 39–42

Sub-theme and content	Using this book
3.1.3.3 Coastal landscape development A variety of coastal landscapes are to be studied and these must include some from outside the United Kingdom but may also include UK examples. The requirement is to study the factors and processes in their origins and development. ■ Landforms and landscapes of coastal erosion: cliffs and wave-cut platforms, features of cliff profiles including caves, arches and stacks ■ Landforms and landscapes of coastal deposition: beaches, spits (simple and compound), tombolos, offshore bars, barrier beaches and islands, sand dunes ■ Estuarine mudflat/salt marsh environments and associated landscapes ■ Eustatic, isostatic and tectonic sea level change including the major changes in sea level of the last 10 000 years ■ Emergent and submergent coastlines: raised beaches, marine platforms, rias, fjords, Dalmatian coasts Study is also to be made of recent and predicted change and the potential impact this change might bring to the coastal zone.	Chapter 3 pages 54–63 Chapter 3 pages 63–77; Chapter 4 pages 95–103 Chapter 3 pages 78–9; Chapter 4 pages 90–5 Chapter 5 pages 113–22 Chapter 5 pages 122–8
3.1.3.4 Coastal management The theme here is on human interventions in coastal landscapes. A contrast can be made between traditional approaches to risks from coastal flood and erosion of both hard and soft engineering and more recent sustainable approaches such as shoreline management and integrated coastal zone management.	Chapter 5 pages 28–37 Chapter 7 pages 176–202
3.1.3.5 Quantitative and qualitative skills A range of quantitative and qualitative skills should be encountered set within a study of landscape systems. These skills should include observation, measurement and geospatial mapping as well as the manipulation and application of statistical skills to data arising from field measurements.	
3.1.3.6 Case studies Students should undertake case study(ies) of coastal environment(s) at a local scale. These should allow fundamental coastal processes to be illustrated and analysed along with their landscape outcomes. The challenges encountered in the sustainable management of the locations are to be considered. This is also an opportunity to engage with field data. Students should undertake a case study of a coastal landscape outside the UK to illustrate and analyse risks and opportunities for human occupation and development. It should also encourage students to evaluate human responses of resilience, mitigation and adaptation.	Chapter 3 page 53, page 69, page 72 Chapter 5 pages 136–7 Chapter 7 pages 180–1, page 189, pages 196–9

AQA assessment guidance

Coastal Systems and Landscapes is assessed as part of Paper 1 (7037/1). This examination is 2 hours and 30 minutes in duration and has a total mark of 120.

There are 36 marks allocated for Coastal Systems and Landscapes. The 36 marks consist of:

- a series of three short-answer questions (worth 16 marks in total)
- one 20-mark evaluative essay.

Short-answer questions (up to 6 marks)

The first question will most likely be a purely knowledge-based short-answer task targeted at AO1 (assessment objective 1) using the command word 'explain'. High marks will be awarded to students who can write concise, detailed answers which incorporate and link together a range of geographical ideas, concepts or theories. As a general rule, try to ensure that every point you make is either developed or exemplified:

- A developed point takes the explanation a step further (perhaps providing additional detail of how a process operates).
- An exemplified point refers to a relatively detailed or real-world example in order to support the explanation with evidence.

The second short-answer question will make use of one or more resources such as maps, graphs or data tables and is targeted at AO3 (assessment objective 3). This means that you will be required to use geographical skills (AO3) to analyse or extract meaningful information or evidence from the figure. These questions will most likely use the command words 'analyse', 'compare' or 'assess'. The 'Analysis and interpretation' features included throughout this book are intended to support the study skills you need to answer this kind of question successfully. There may be the need to carry out some mathematical operation such as completing a statistical test and then interpreting its result.

The third and final short-answer question will again make use of a resource such as a photograph or diagram but is now targeted mainly at AO2 (assessment objective 2), as well as offering some AO1 marks. It may ask you to 'assess' something about a particular process in the context of the location in the resource. It also will ask you to use 'your own knowledge'. This means you are expected to use the resource only as a 'springboard' to apply your own geographical ideas and information. For example, a 6-mark question accompanying a photograph of a sand dune system might ask: 'Using the figure and your own knowledge, assess the role of transport in the development of sand dunes'. You can answer by writing about your own case studies and the benefits these places have gained.

Evaluative essay writing

The 20-mark essay will most likely use a command word or phrase such as 'how far', 'to what extent' or 'discuss'. The mark scheme is weighted equally towards AO1 and AO2. For instance:

How far is it possible to prevent coastal flooding?

'Coastal landscapes are as much the product of past processes as present-day ones.' Discuss.

Every chapter of this book contains a section called 'Evaluating the issue'. These have been designed specifically to support the development of the evaluative essay writing skills you need to tackle tough titles such as these. As you read each 'Evaluating the issue' section, pay particular attention to the following:

- *Underlying assumptions and possible contexts should be identified at the outset* Considering the different sources of coastal flooding is helpful, such as particularly high tides, storm surges or tsunami. In the second question given above, a phrase such as 'past processes' could be explored a little by considering various time scales such as geological time (millions of years) or the period since the last major ice advance (thousands of years). A question such as 'what processes were operating in the past?' is relevant.

- *An essay needs to be carefully structured around different themes, views, scales, topic connections or arguments* In the first essay, one possible approach might be to look at different examples to exemplify different approaches. Comparing the hard engineering of the Dutch Delta scheme with attempts to manage a coastline with more natural techniques could be valuable. For essays which ask you to evaluate a viewpoint or quotation, such as the one shown above, answers which score highly are likely to be well-balanced insofar as roughly half of the main body of the essay will consider ideas and arguments supporting the statement, and the remaining half will deal with counter-arguments.

- *The command word 'evaluate' requires you to reach a final judgement* Don't just sit on the fence. Draw on all the arguments and facts you have already presented in the main body of the essay, weigh up the entirety of your evidence and say whether – on balance – you agree or disagree with the question you were asked. To guide you, here are three simple rules.

  1 *Never sit on the fence completely.* The essay titles have been created purposely to generate a discussion which invites a final judgement following debate. Do not expect to receive a high mark if you end your essay with a phrase such as: 'It is possible to prevent coastal flooding but at some times it is not.'

  2 *Equally, it is best to avoid extreme agreement or disagreement.* In particular, you should not begin your essay by dismissing one viewpoint entirely, for example by writing: 'In my view, it is impossible to prevent coastal flooding and this essay will explain all of the reasons why.' Instead, you should consider different points of view.

  3 *An 'agree, but...' or 'disagree, but...' judgement is usually the best position to take.* This is a mature viewpoint which demonstrates you are able to take a stand on an issue while remaining mindful of other views and perspectives.

Synoptic geography

In addition to the three main AOs, some marks are awarded for 'synopticity'. Instead of focusing on one isolated topic, you are expected to draw together information and ideas from across the specification in order to make connections between different 'domains' of knowledge, especially links between people and the environment (i.e. connections across human geography and physical geography). The study of challenges facing communities living in the Maldives (page 139) and around the Exe estuary (page 196) is a good example of synoptic geography because of the important linkages between rising sea level and placemaking.

Throughout your course, take careful note of synoptic themes whenever they emerge in teaching, learning and reading. Examples of synoptic themes could include: how movements of water through the landscape affect the shape and development of cliffs; how coastal management schemes might alter the characteristics of places; how globalisation of the tourism industry can affect coastal ecosystems such as coral reefs and sand dune systems.

Whenever you finish reading a chapter in this book, make a careful note of any synoptic themes that have emerged (they may have been highlighted, or these could be linkages that you work out for yourself).

AQA's synoptic assessment

Some 9-mark or 20-mark exam questions may require you to link together knowledge and ideas from different topics you have learned about. These may appear in both your physical geography and human geography examination papers. For example:

- An ecosystem question (Paper 1) might ask you to think about ways in which rising sea level could affect marine ecosystems.
- A resource issue question (Paper 2) could ask you to discuss ways in which energy security might be influenced by reappraising coastal energy sources such as tides and waves.

In the exam, your Coastal Systems and Landscape essay might include a synoptic link. For example, the following question:

Assess the extent to which sustainable use of coastal landscapes can be balanced with rising demand for coastal locations by human activities.

The mark scheme would include the following statement: 'This question requires links to be made across the specification, specifically between how coastal landscapes might be managed sustainably and the need for space for human activities such as housing, industry, transport infrastructure and food production'. One way to tackle this kind of potentially tricky question is to draw a concept map when planning your response. Draw two equally-sized circles and fill these with relevant ideas, processes and contexts, trying to achieve the best balance you can between the two linked topics.

Pearson Edexcel: Coastal Landscapes and Change

Content guidance

The optional topic of Coastal Landscapes and Change is supported fully by this book.

The Coastal Landscapes and Change option deals with the operation of the coastal system to produce distinctive coastal landscapes. This is achieved through the interaction of energy flows, such as winds and waves, sediment flows and the influence of geology. Landscapes such as rocky, sandy and estuarine develop and are increasingly threatened by both physical processes and human activities. There is, therefore, a need for holistic and sustainable management of coastal locations.

The study of these processes and patterns must include examples of landscapes from inside and outside the UK.

Coastal Landscapes and Change (Option 2B)

The content is structured around four main enquiry questions.

Enquiry question and content	Using this book
1 Why are coastal landscapes different and what processes cause these differences?	
The topic begins by defining the location of the coast (littoral zone) and establishing a basic contrast between rocky (high energy) and sandy/estuarine (low energy) coasts. The role of geological structure is considered both on coastal plans and profiles such as concordant/discordant coasts and cliffs. The relative stability or otherwise of coasts, both rocky and sandy/estuarine, is explored using factors such as geology and vegetation.	Chapter 1 pages 1–3 Chapter 3 pages 51–89 Chapter 4 pages 90–103
2 How do characteristic coastal landforms contribute to coastal landscapes?	
The importance of marine erosion due to the interaction of wave energy and geology on rocky landforms is studied, as is the role of wave energy on beach morphology. Sediment transport and deposition is a focus set in the context of a variety of landforms. The concept of the sediment cell with feedback, both negative and positive, operating within it is to be studied. You should also be familiar with the idea of dynamic equilibrium in this context.	Chapter 1 pages 3–22, Chapter 2 pages 28–45 Chapter 2 pages 36–45 Chapter 2 pages 31–6
The role of sub-aerial processes is an important factor in coastal landform development, in particular cliff profiles.	
3 How do coastal erosion and sea level change alter the physical characteristics of coastlines and increase risks?	Chapter 5 pages 113–28
Coastal landforms and landscapes change over the course of different time scales. Over the longer term, the complex interaction of eustatic, isostatic and tectonic factors leads to the development of emergent and submergent coastlines with characteristic features, such as raised beaches and rias. Contemporary sea level change arises from the effects of global warming as well as tectonic adjustments.	Chapter 5 pages 132–7 Chapter 7 pages 176–81
The study of how rapid coastal retreat poses risks to people includes understanding how both physical and human factors can be influences. It is important to appreciate that rates of recession are not constant and that extreme events can prove especially damaging both to the coastline and to human activities. In addition, the threats posed by coastal flooding are to be considered, with causal factors such as storm surges and sea level rise associated with climate change investigated.	Chapter 5 pages 128–37

Enquiry question and content	Using this book
4 How can coastlines be managed to meet the needs of all players?	Chapter 5 pages 128–37
The consequences of the impacts of receding coastlines and coastal flooding on communities is the focus in this section. Consideration is to be given both to economic and social losses in coastal communities in countries at various points along the development continuum. The prospect of environmental refugees forced out of coastal locations is relevant.	
In the light of the risks coastal locations pose, various approaches to managing those risks are to be studied. Both hard and soft engineering schemes are to be assessed in terms of their costs and benefits.	Chapter 7 pages 181–205
Students are to investigate the widespread use of holistic and sustainable management schemes. The various decisions as to what approach different stretches of coastline receive are to be studied and the reasons why some locations are fully defended while others are not investigated. This then leads naturally into considering the conflicts, potential and real, that can arise among different players involved in the coastal zone.	

Pearson Edexcel assessment guidance

Coastal Landscapes and Change is assessed as part of Paper 1 (9GE0/01). This examination is 2 hours and 15 minutes in duration and has a total mark of 105. There are 40 marks allocated for the Coastal Landscapes and Change question. The 40 marks consist of:

- three shorter-length questions allocated 6, 6 and 8 marks respectively
- one 20-mark evaluative essay.

Shorter-answer questions

You will encounter three shorter-length questions:

- The two 6-mark questions are targeted equally at AOs 1 and 2 (assessment objectives 1 and 2) and questions will be linked to a resource such as a photograph or diagram and typically use the command word 'explain'. For example, a 6-mark question accompanying a photograph might ask: 'Study the figure. Explain how the wind has contributed to the formation of the sand dunes shown'. To score full marks, you must (i) apply geographical knowledge and understanding to this new context you are being shown, that is the actual image, and (ii) establish very clear connections between the focus of the question that is being asked (wind) and the feature (sand dunes) you have been shown.
- The 'Analysis and interpretation' features included throughout this book are intended to support the study skills you need to answer this kind of question successfully.

You will also be asked one purely knowledge-based question targeted at AO1 (assessment objective 1). This will most likely be the third shorter-answer question and use the command word 'explain', such as 'Explain how sea level change contributes to the understanding of coastal landscapes'. High marks will be awarded to students who can write concise, detailed answers which incorporate and link together a range of geographical ideas, concepts or theories. As a general rule, try to ensure that every point you make is either developed or exemplified:

- A developed point takes the explanation a step further (perhaps providing additional detail of how a process operates).
- An exemplified point refers to a relatively detailed or real-world example in order to support the explanation with evidence.

Evaluative essay writing

The 20-mark essay will most likely use the command word 'evaluate'. The mark scheme is weighted heavily towards AO2. For instance:

Evaluate the advantages of adopting a "hold the line" approach to coastal management.

Evaluate the role of sub-aerial processes in the development of cliff profiles.

Evaluate the value of a holistic approach to dealing with risks from coastal flooding.

Every chapter of this book contains a section called 'Evaluating the issue'. These have been designed specifically to support the development of evaluative essay writing skills. As you read each 'Evaluating the issue' section, pay particular attention to the following:

- *Underlying assumptions and possible contexts should be identified at the outset* Considering the different sources of coastal flooding is helpful, such as particularly high tides, storm surges or tsunami. In the third question given above, a word such as 'risks' should be explored, for example by considering economic, social and environmental risks arising from coastal flooding. Before planning your answer think very carefully about what could be considered as 'costs' and 'benefits' so that more than simply a monetary aspect is considered, for example the importance of a beach as an amenity.
- *An essay needs to be carefully structured around different themes, views, scales, topic connections or arguments* In the first essay, what other approaches to coastal management should be considered? In the second essay, what else could be important other than sub-aerial processes? For essays which ask you to evaluate a viewpoint, answers which score highly are likely to be well-balanced insofar as roughly half of the main body of the essay will consider ideas and arguments supporting the statement, and the remaining half will deal with counter-arguments.
- *The command word 'evaluate' requires you to reach a final judgement* Don't just sit on the fence. Draw on all the arguments and facts you have already presented in the main body of the essay, weigh up the entirety of your evidence and say whether – on balance – you agree or disagree with the question you were asked. To guide you, here are three simple rules.

  1 *Never sit on the fence completely.* The essay titles have been created purposely to generate a discussion which invites a final judgement following debate. Do not expect to receive a high mark if you end your essay with a phrase such as: 'So all in all, a hold the line policy has both advantages and disadvantages.'

  2 *Equally, it is best to avoid extreme agreement or disagreement.* In particular, you should not begin your essay by dismissing one viewpoint entirely, for example by writing: 'In my view, sub-aerial processes are the most significant factor in the development of cliff profiles and this essay will explain all of the reasons why'. It is essential that you consider different points of view.

  3 *An 'agree, but...' or 'disagree, but...' judgement is usually the best position to take.* This is a mature viewpoint which demonstrates you are able to take a stand on an issue while remaining mindful of other views and perspectives.

Synoptic geography

In addition to the three main AOs, some of your marks are awarded for 'synopticity'. Instead of focusing on one isolated topic, you are expected to draw together information and ideas from across your specification in order to make connections between different 'domains' of knowledge, especially links between people and the environment (i.e. connections across human geography and physical geography).

Throughout your course, take careful note of synoptic themes whenever they emerge in teaching, learning and reading. Examples of synoptic themes could include: how movements of water through the landscape affect the shape and development of cliffs; how coastal management schemes might alter the characteristics of places; how globalisation of the tourism industry can affect coastal ecosystems such as coral reefs and sand dune systems.

Whenever you finish reading a chapter in this book, make a careful note of any synoptic themes that have emerged (they may have been highlighted, or these could be linkages that you work out for yourself).

Pearson Edexcel's synoptic assessment

Synoptic exam questions are worth plenty of marks and you need to be well-prepared for them.

- In the Pearson Edexcel course, an entire examination paper is devoted to synopticity: Paper 3 (2 hours 15 minutes) is a synoptic 'decision-making' investigation. It consists of an extended series of data analysis, short-answer tasks and evaluative essays (based on a previously unseen resource booklet).
- As part of your Paper 3 answers, you will need to apply a range of knowledge from different topics you have learned about and also make good analytical use of the previously unseen resource booklet (the 'Analysis and interpretation' features in this book have been carefully designed to help you in this respect). The context used in the resource booklet may well make use of themes drawn from Coastal Landscapes and Change, though for the sake of fairness it must be equally relevant to Option 2A Glaciated Landscapes and Change. For example, the diverging attitudes of different players about attempts to manage a fragile environment lend themselves particularly well to this kind of synoptic assessment.

③ OCR: Coastal Landscapes

Content guidance

The optional topic of Coastal Landscapes is supported fully by this book.

The Coastal Landscapes topic introduces students to the integrated study of surface processes, landforms and resultant landscapes using the conceptual framework of a systems approach. You should develop knowledge and understanding of Earth surface processes such as weathering, erosion and mass movement in a coastal context. Associated with this is a study of the transfers of energy (for example, waves) and flows of materials (for example, sediment) which occur within coastal landscapes. Individual landforms are formed and evolve in response to these transfers and flows. Coastal landscapes form from the association and interaction of characteristic landforms and both landforms and landscapes are dynamic in time and space. The topic also allows you to investigate the interactions and connectivity between human activities and coastal environments.

Coastal Landscapes (Topic 1.1 Option A)

This section of the OCR specification explores how the coastal landscape can be viewed as a system, how landforms develop within the coastal zone, and the influences of both climate and human activities on this landscape system. The detailed content is structured around four sub-themes.

Enquiry question and content	Using this book
1 How can coastal landscapes be viewed as a system? This section establishes a conceptual overview of the various components of the coastal landscape system. Aspects such as inputs, stores/components, flows/transfers and outputs are to be considered. Concepts such as feedback (positive/negative) and equilibrium are important when thinking about flows of energy and materials. The role of winds, waves and tides as energy sources, the geological input and various sources of sediment are to be considered.	Chapter 1 pages 1–27 Chapter 2 pages 36–8
2 How are coastal landforms developed? An understanding of how flows of energy and materials influence geomorphic processes is required so that you appreciate how coastal landforms develop. Two groups of landforms are identified, one being landforms predominantly influenced by erosion (higher energy), the other predominantly by deposition (lower energy). It is important to appreciate that coastal landforms are often interrelated and combine to make up characteristic coastal landscapes. Two types should be identified and studied, one high energy and the other low energy, such as rocky and estuarine locations respectively. Consideration should also be given to how and why landscape systems change over time and that the time periods over which change occurs ranges from the short term (cliff fall in seconds), through medium term (seasonal change in beach profile) to the long term (raised beach uplift over millennia).	Chapter 2 pages 28–50 Chapter 3 pages 51–79 Chapter 4 pages 90–5
3 How do coastal landforms evolve over time as climate changes? The role of longer-term climate change and consequent changes in sea level and the implications for coastal landform development is to be studied. Investigation of both emerging and submerging coastal landscapes should be undertaken. You should also consider the modifications of the landforms by present and future climate and sea level change.	Chapter 5 pages 113–28
4 How does human activity cause change within coastal landscape systems? In this section the focus is on the interaction, intentional or unintentional, between human activity and coastal landscape systems. You should develop two case studies that exemplify the effects management or unintentional change (for example, port or resort development) can have on components of the coastal system, such as processes and flows of material and energy. The effects of management strategies or economic developments on coastal landforms should be studied, such as changes in the morphology of the landforms, as well as any consequences for the coastal landscape.	Chapter 5 pages 132–41 Chapter 6 pages 146–8 Chapter 7 pages 176–206

OCR assessment guidance

Coastal Landscapes is assessed as part of Paper 1 (H481/01). This examination is 1 hours and 30 minutes in duration and has a total mark of 66. There are 33 marks allocated for Coastal Landscapes, indicating that you should spend around 45 minutes answering it. The 33 marks consist of:

- a series of three or more short-answer questions worth 9 marks in total
- a medium-length question worth 8 marks
- one 16-mark evaluative essay.

Short and medium-length questions

Two of the three short-answer questions are targeted at AO3 (assessment objective 3), the third at AO2 (assessment objective 2). In those targeting AO3 there may be the need to carry out some mathematical operation such as completing a statistical test analysing some data. It is also likely that you will be required to use geographical skills (AO3) to extract meaningful information or evidence from a resource, such as a photograph or chart.

- These questions are likely to use the command words 'calculate' and 'explain'. They will also include the instruction 'With reference to Figure 1' or 'Study Figure 1'.
- The 'Analysis and interpretation' features included in all chapters of this book are intended to support the study skills you need to answer this kind of question successfully.

The medium-length question worth 8 marks targets AO1 (assessment objective 1). This is a standalone question allowing you to use your knowledge and understanding of coastal processes and landforms to set out the causes of a phenomenon and/or the factors influencing its nature. The command word 'explain' could be used for this question.

High marks will be awarded to students who can write concise, detailed answers which incorporate and link together a range of geographical ideas, concepts or theories. As a general rule, try to ensure that every point you make is either developed or exemplified:

- A developed point takes the explanation a step further (perhaps providing additional detail of how a process operates).
- An exemplified point refers to a relatively detailed or real-world example in order to support the explanation with evidence.

Evaluative essay writing

The 16-mark essay will most likely use a command word or phrase such as 'how far do you agree', 'to what extent' or 'discuss'. The mark scheme is weighted equally towards AO1 and AO2. For instance:

'Wave energy is the most significant input to the coastal system.' To what extent do you agree with this statement?

How far do you agree that the consequences of coastal management strategies have positive outcomes?

Every chapter of this book contains a section called 'Evaluating the issue'. These have been designed specifically to support the development of evaluative essay writing skills. As you read each 'Evaluating the issue' section, pay particular attention to the following:

- *Underlying assumptions and possible contexts should be identified at the outset* For the first question shown above, before planning your answer think very carefully about the different inputs the coastal system receives. This is why a secure knowledge and understanding of the coastal system is important. In the second question, consider the range of management strategies that are deployed in the coastal zone. Important parameters such as these should be established at the planning stage of your essay and may be mentioned in the introduction.
- *An essay needs to be carefully structured around different themes, views, scales, topic connections or arguments* In the second essay, what different kinds of strategy are there, and how and why might these strategies differ in outcome? In the first essay, can you come up with counter-arguments showing that wave energy may not be the most significant input in all locations and at all times? For essays which ask you to discuss or evaluate how far you agree with a viewpoint, answers which score highest are likely to be well-balanced insofar as roughly half of the main body of the essay will consider ideas and arguments supporting the statement, and the remaining half will deal with counter-arguments.
- *Command words and phrases such as 'to what extent' and 'discuss' require you to reach a final judgement* Don't just sit on the fence. Draw on all the arguments and facts you have already presented in the main body of the essay, weigh up the entirety of your evidence and say whether – on balance – you agree or disagree with the question you were asked. To guide you, here are three simple rules.

1 *Never sit on the fence completely.* Essay titles are created purposely to generate a discussion which invites a final judgement following debate. Do not expect to receive a really high mark if you end an essay with a phrase such as: 'So all in all, wave energy is the most significant input to the coastal system'.
2 *Equally, it is best to avoid extreme agreement or disagreement.* In particular, you should not begin your essay by dismissing one viewpoint entirely, for example by writing: 'In my view, coastal management strategies have nothing but negative outcomes and this essay will explain all of the reasons why'. It is essential that you consider different points of view.
3 *An 'agree, but...' or 'disagree, but...' judgement is usually the best position to take.* This is a mature viewpoint which demonstrates you are able to take a stand on an issue while remaining mindful of other views and perspectives.

Synoptic geography

In addition to the three main AOs, some of your marks are awarded for 'synopticity'. Instead of focusing on one isolated topic, you are expected to draw together information and ideas from across your specification in order to make connections between different 'domains' of knowledge, especially links between people and the environment (i.e. connections across human geography and physical geography). The study of challenges facing communities living in the Maldives (page 136) and around the Exe estuary (page 196) is a good example of synoptic geography because of the important linkages between rising sea level and placemaking.

Throughout your course, take careful note of synoptic themes whenever they emerge in teaching, learning and reading. Examples of synoptic themes could include: how movements of water through the landscape affect the shape and development of cliffs; how coastal management schemes might alter the characteristics of places; how globalisation of the tourism industry can affect coastal ecosystems such as coral reefs and sand dune systems.

Whenever you finish reading a chapter in this book, make a careful note of any synoptic themes that have emerged (they may have been highlighted, or these could be linkages that you work out for yourself).

OCR's synoptic assessment

In the OCR course, part of Paper 3 (Geographical Debates H481/03) is devoted to synopticity. For this exam, you will have studied two optional topics chosen from: Climate Change, Disease Dilemmas, Exploring Oceans, Future of Food and Hazardous Earth. In Section B of Paper 3, you must answer two synoptic essays worth 12 marks each (in total, this adds up to 24 marks).

Each synoptic essay links together the chosen option with a topic from the core of the A-level course. Possible Paper 3 essay titles might therefore include:

Examine how climate change may be impacting coastal landforms.

'Short-term adaptations to rising sea level are no more than a "sticking plaster" solution for island communities.' How far do you agree with the statement?

One way to tackle these kind of questions would be to draw a concept map to help plan your response. Draw two equally-sized circles and fill these with relevant ideas, processes and contexts, trying to achieve the best balance you can between the two linked topics. The mark scheme requires that your answer includes: 'clear and explicit attempts to make appropriate synoptic links between content from different parts of the course of study'.

④ WJEC and Eduqas: Changing Landscapes

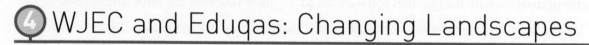

Content guidance

The study of the optional theme of Coastal Landscapes within the Changing Landscapes section is supported fully by this book.

The Coastal Landscapes theme focuses on the dynamic nature of physical systems and processes and on the interactions and connectivity between people, places and environments in both time and space. Study should focus on:

- the interaction of winds, waves and currents
- supply of sediment from terrestrial and offshore sources
- spatial and temporal variations in geomorphological processes
- how flows of energy and materials combine to create specific landforms on rocky, sandy and estuarine coastlines.

Coastal Landscapes (Theme 1.1)

This section of both the WJEC and Eduqas specifications is structured around ten short sub-themes.

Sub-theme and content	Using this book
1.1.1 The operation of the coast as a system This section establishes a conceptual overview of the various components of the coastal landscape system, including inputs, stores and transfers of energy and materials and outputs.	Chapter 1 pages 1–5
It includes supplies of sediment (offshore and terrestrial) as well as the concept of the sediment cell.	Chapter 1 pages 19–22
The idea of dynamic equilibrium in the coastal system is to be studied.	
1.1.2 Temporal variations and their influence on coastal environments The energy inputs of different types of waves are to be considered, along with diurnal tides and offshore and onshore currents.	Chapter 1 pages 5–18
1.1.3 Landforms and landscape systems, their distinctive features and distribution In this section two contrasting types of coastal environment are studied: ● High energy, rocky coastlines with their erosional landforms and landscape systems ● Low energy, sandy and estuarine coastlines with their depositional landforms and landscape systems.	Chapter 3 pages 28–86
1.1.4 Factors affecting coastal processes and landforms The fetch, wave type, orientation of waves, wave refraction and reflection are investigated. The geological factors of structure (bedding, dip, joints, folding and faulting) and lithology (mineral composition, hardness and solubility) of rocks are to be considered.	Chapter 1 pages 8–14 Chapter 3 pages 51–63
1.1.5 Processes of coastal weathering, mass movement, erosion and the characteristics and formation of associated landforms and landscapes Here the diversity of sub-aerial weathering processes (physical, chemical, biotic) and mass movements including landslides, slumps and rock falls are to be studied.	Chapter 2 pages 31–36
Marine erosional processes to be studied are hydraulic action, abrasion (corrasion), corrosion and attrition.	Chapter 2 pages 29–31
The study of characteristics of coastal landforms and landscapes both for and beyond the UK includes cliffs, bays and headlands, cave-arch-stack-stump sequence, wave-cut platforms, geos and blowholes.	Chapter 3 pages 51–63
1.1.6 Processes of coastal transport and deposition and the characteristics and formation of associated landforms and landscapes Here the range of transport processes, solution, suspension, saltation and traction including longshore drift are to be studied.	Chapter 2 pages 36–42
Coastal depositional processes including flocculation and sediment sorting brought about by energy reductions are to be considered.	Chapter 4 pages 92–3
The study of characteristics of coastal landforms and landscapes both for and beyond the UK includes beaches, spits, bars, tombolos and cuspate forelands.	Chapter 3 pages 63–77

Sub-theme and content	Using this book
1.1.7 Aeolian, fluvial and biotic processes and the characteristics and formation of landforms in coastal environments	
Wind action and associated landforms of sand dunes are to be considered in this section.	Chapter 4 pages 95–103
Estuarine environments and their associated landforms of tidal flats and salt marsh, including micro-features such as channels and rills are also to be studied.	Chapter 3 pages 78–91 Chapter 4 pages 104–7
Biotic processes are to be considered in the context of coral reefs and mangrove.	
1.1.8 Variations in coastal processes, coastal landforms and landscapes over different timescales	
It is important to appreciate change in the coastal zone across different timescales.	Chapter 3 pages 54–9
Short-term changes (seconds and minutes) due to high energy storms and rapid mass movements bring about alterations in cliff profiles, for example.	Chapter 3 pages 75–6
Beaches can alter seasonally in their profiles as wave energy varies.	Chapter 5 pages 115–6
Eustatic and isostatic changes occur over millennia and alter landforms.	
1.1.9 Coastal processes are a vital context for human activity	
Consideration is to be given to the positive and negative impacts of coastal processes on human activity.	Chapter 4 pages 106–7 Chapter 5 pages 129–30
Positive impacts such as tourism and negative such as economic and social losses associated with coastal erosion are to be studied.	Chapter 6 pages 146–8, 156–62
A case study of a management strategy to manage impacts of coastal processes on human activity should be developed.	Chapter 5 pages 136–7
1.1.10 The impact of human activity on coastal landscape systems	
In this section it is the positive and negative impacts of human activity on coastal processes and landforms that are the focus of study.	Chapter 6 pages 154–62 Chapter 7 pages 176–205
Positive impacts such as management and conservation and negative impacts such as offshore dredging and erosion of sand dunes are relevant examples.	
A case study of a management strategy to manage impacts of human activity on coastal processes and landforms should be developed.	

WJEC and Eduqas assessment guidance

Coastal Landscapes is assessed as part of:

- *WJEC Unit 1* This examination is 2 hours long, and has a total mark of 64. There are 32 marks allocated for Coastal Landscapes, indicating that you should spend around 45 minutes answering. The 32 marks consist of:
  - a series of short-answer questions worth between 2 and 8 marks
  - 8-mark questions that are best considered as 'mini-essays'
  - a few marks allocated to skills such as a simple calculation or completion of a diagram such as a graph, using data provided.
- *Eduqas Component 1* This examination is 1 hour and 45 minutes in duration, and has a total mark of 82. There are 41 marks allocated for Coastal Landscapes, indicating that you should spend around 45 minutes answering. The 41 marks consist of:

- a series of short-answer questions (worth 26 marks in total)
- one 15-mark evaluative essay (from a choice of two).

Both courses use broadly similar assessment models and these are dealt with jointly below.

Short-answer questions

Some of the questions you are asked will be linked to resources (maps, charts, tables or photographs).

- The opening question(s) will be targeted at AO3 (assessment objective 3). This means that you will be required to use geographical skills (AO3) to analyse or extract meaningful information or evidence from the resource. These questions will most likely use command words including 'examine', 'analyse' or 'compare'. The 'Analysis and interpretation' features included throughout this book are intended to support the study skills you need to answer this kind of question successfully.
- Alternatively, you could be required to briefly complete a short skills-based numerical or graphical task. Your specification includes a list of skills and techniques you're expected to be able to carry out, such as a Spearman statistical test, the calculation of an interquartile range or accurate plotting of data on to a chart or graph.
- Finally, you may be asked to give a possible explanation of the information shown in the figure using applied knowledge. This kind of question is targeted at AO2 (assessment objective 2) and will most likely use the command word 'suggest'. It will also include the instruction: 'With reference to Figure 1'.
- For example, a series of short questions could accompany an Ordnance Survey map showing a stretch of the coastal zone. The opening (AO3) question could be: 'Compare the coastal landforms shown in Figure 1'. The AO2 question which follows might ask: 'Suggest reasons why the coastal landforms vary along the stretch of coastline shown in Figure 1.' To score full marks, you must (i) apply geographical knowledge and understanding to this new context you are being shown, and (ii) establish very clear connections between the question that is being asked and the stimulus material you have been shown.

Eduqas

In the Eduqas examination, some standalone short-answer questions are not accompanied by a figure. They are purely knowledge-based, targeted at AO1 (assessment objective 1) and worth up to 6 marks. They will most likely use the command word 'explain', 'describe' or 'outline'. For example: 'Describe two ways in which coastal sediment is transported.' High marks will be awarded to students who can write concise, detailed answers which incorporate and link together a range of geographical ideas, concepts or theories. As a general rule, try to ensure that every point you make is either developed or exemplified:

- A developed point takes the explanation a step further (perhaps providing additional detail of how a process operates).
- An exemplified point refers to a relatively detailed or real-world example in order to support the explanation with evidence.

WJEC

In the WJEC examination, some short-answer questions are not accompanied by a figure. They are targeted jointly at AO1 (assessment objective 1) and AO2 (assessment objective 2), and are worth 8 marks. They will most likely use the command words 'describe and explain', 'examine' or 'assess'. For example: 'Describe and explain how changes in sea level result in the formation of one coastal landform.' High marks will be awarded to students who can write concise, detailed answers which incorporate and

link together a range of geographical ideas, concepts or theories. As a general rule, try to ensure that every point you make is either developed or exemplified:

- A developed point takes the explanation a step further (perhaps providing additional detail of how a process operates).
- An exemplified point refers to a relatively detailed or real-world example in order to support the explanation with evidence.

Evaluative essay writing (Eduqas only)

The 15-mark essays (10 marks AO1, 5 marks AO2) which feature in Eduqas Component 1 may use command words such as 'assess' and 'examine'. For instance:

Examine the importance of mass movement in the development of one coastal landform.

Assess the varied impacts of human activity on coastal processes and landforms.

Every chapter of this book contains a section called 'Evaluating the issue'. These have been designed specifically to support the development of evaluative essay writing skills. As you read each 'Evaluating the issue' section, pay particular attention to the following:

- *Underlying assumptions and possible contexts should be identified at the outset* For the first question shown above, before planning your answer think very carefully about the types of mass movement that are common around the coast and the range of other factors involved in landform development.
- *An essay needs to be carefully structured around different themes, views or scales of analysis* In the second essay, think carefully about possible human activities the essay could be structured around. Which different aspects of human activities might be relevant, such as management of sediment movement, conservation of sand dunes or development of seaside resorts? Important parameters such as these should be established at the planning stage of your essay and may be included in an introduction.

Synoptic geography

In addition to the three main AOs, some of your marks are awarded for 'synopticity'. Instead of focusing on one isolated topic, you are expected to draw together information and ideas from across your specification in order to make connections between different 'domains' of knowledge, especially links between people and the environment (i.e. connections across human geography and physical geography). The study of challenges facing communities living in the Maldives (page 136) and around the Exe estuary (page 196) is a good example of synoptic geography because of the important linkages between rising sea level and placemaking.

Throughout your course, take careful note of synoptic themes whenever they emerge in teaching, learning and reading. Examples of synoptic themes could include: how movements of water through the landscape affect the shape and development of cliffs; how coastal management schemes might alter the characteristics of places; how tectonic processes can affect coastal landforms.

Whenever you finish reading a chapter in this book, make a careful note of any synoptic themes that have emerged (they may have been highlighted, or these could be linkages that you work out for yourself).

The WJEC and Eduqas synoptic assessment

In the WJEC course, part of Unit 3 is devoted to synopticity, while for Eduqas a similar assessment appears in Component 2. In both cases, synopticity is examined using an assessment called '21st Century Challenges'. This synoptic exercise consists of a linked series of four figures (maps, charts or photographs) with a choice of two accompanying essay questions. The WJEC question has a maximum mark of 26; for Eduqas it is 30. Possible questions include:

Discuss the severity of the different risks that cities increasingly face.

To what extent could the management of different risks lead to changes in the characteristics of urban places?

As part of your answer, you will need to apply a range of knowledge from different topics you have learned about, and also make good analytical use of the previously unseen resources in order to gain AO3 credit (the 'Analysis and interpretation' features in this book have been carefully designed to help you in this respect). One or both of the questions may relate quite clearly to the topic of Coastal Landscapes, as the example titles above demonstrate:

- Risks to cities could include the threats to coastal cities from rising sea level and increasing severity of storms.
- Note also how the second question allows you to explore not just 'real world' changes to infrastructure such as an increased scale of coastal defences which may involve a loss in amenity value, but also changing place perceptions (linked with feelings of increased danger or vulnerability) linked to increased wave energy and coastal flood risks.

Command words and phrases such as 'to what extent' and 'discuss' require you to reach a final judgement. Don't just sit on the fence. Draw on all the arguments and facts you have already presented in the main body of the essay, weigh up the entirety of your evidence and say whether – on balance – you agree or disagree with the question you were asked. To guide you, here are three simple rules.

1 *Never sit on the fence completely.* Essay titles are created purposely to generate a discussion which invites a final judgement following debate. Do not expect to receive a really high mark if you end an essay with a phrase such as: 'So all in all, there are many risks which cities face, and they are all important'.

2 *Equally, it is best to avoid extreme agreement or disagreement.* In particular, you should not begin your essay by dismissing one viewpoint entirely, for example by writing: 'In my view, climate change is the greatest risk that all places face and this essay will explain all of the reasons why this is the case.' It is essential that you consider a range of arguments or different points of view.

3 *An 'agree, but...' or 'disagree, but...' judgement is usually the best position to take.* This is a mature viewpoint which demonstrates you are able to take a stand on an issue while remaining mindful of other views and perspectives.

Index

Acknowledgements

Text permissions

p.107 Figure 4.17 Reprinted with permission of ARC Centre of Excellence for Coral Reef Studies at James Cook University; **p.121** Figure 5.7 FAQ 5.1, Figure 1 from Climate Change 2007: The Physical Science Basis. Working Group I Contribution to the Fourth Assessment Report of the Intergovernmental Panel on Climate Change [Solomon, S., D. Qin, M. Manning, Z. Chen, M. Marquis, K.B. Averyt, M. Tignor and H.L. Miller (eds.)]. Cambridge University Press, Cambridge, United Kingdom and New York, NY, USA; **p.197** Figure 7.16 © SCOPAC Sediment Transport Study. Reprinted with permission.

Photo credits

p.4 © Glebstock - stock.adobe.com; **p.8** Public domain/https://www.metmuseum.org/art/collection/search/45434; **p.10** *l, c, r* © Peter Stiff; **p.13** © Julius Fekete - stock.adobe.com; **p.29** © sergojpg - stock.adobe.com; **p.32, 34** *tl, bl, br,* **37, 39, 53, 55** *t, b,* **56, 57** *l, r,* **60** *tl, tc, tr, b,* **62, 64** © Peter Stiff; **p.72** © SKYSCAN/SCIENCE PHOTO LIBRARY; **p.74, 77, 78, 92, 93** © Peter Stiff; **p.98** © Michael Raw; **p.100** © Peter Stiff; **p.104** © paylessimages - stock.adobe.com; **p.106** © whitcomberd - stock.adobe.com; **p.123** © sophiahilmar - stock.adobe.com; **p.124** © reisegraf.ch - stock.adobe.com; **p.125** © Peter Stiff; **p.126** © Jane Buekett; **p.130** © Copyright Mike Page, All Rights Reserved; **p.135** © Bikeworldtravel - stock.adobe.com; **p.146, 151** *l, r* © Peter Stiff; **p.154** © age fotostock/Alamy Stock Photo; **p.157** © Michael Roper/Alamy Stock Photo; **p.158** © Simon Dawson/Bloomberg via Getty Images; **p.159** © Peter Stiff; **p.165** © East News sp. z o.o./Alamy Stock Photo; **p.180, 183, 184, 185** *t, b* © Peter Stiff; **p.188** © Rijkswaterstaat/Jurriaan Brobbel; **p.196** © Peter Stiff; **p.201** Google Earth/adapted from https://www.sciencedirect.com/science/article/pii/S0169555X15301719/from Geomorphology Volume 256, 1 March 2016, Pages 17-35/https://creativecommons.org/licenses/by/4.0/.